Nucleic acid hybridisation

a practical approach

TITLES PUBLISHED IN
THE
PRACTICAL APPROACH
SERIES

Series editors:
Dr D Rickwood
Department of Biology, University of Essex
Wivenhoe Park, Colchester, Essex C04 3SQ, UK
Dr B D Hames
Department of Biochemistry, University of Leeds
Leeds LS2 9JT, UK

Affinity chromatography

Animal cell culture

Antibodies

Biochemical toxicology

Biological membranes

Carbohydrate analysis

Centrifugation (2nd Edition)

DNA cloning

Drosophila

Electron microscopy
in molecular biology

Gel electrophoresis of nucleic acids

Gel electrophoresis of proteins

HPLC of small molecules

Human cytogenetics

Human genetic diseases

Immobilised cells and enzymes

Iodinated density gradient media

Lymphocytes

Lymphokines and interferons

Mammalian development

Microcomputers in biology

Mitochondria

Mutagenicity testing

Neurochemistry

Nucleic acid and
protein sequence analysis

Nucleic acid hybridisation

Oligonucleotide synthesis

Photosynthesis:
energy transduction

Plant cell culture

Plasmids

Prostaglandins
and related substances

Spectrophotometry
and spectrofluorimetry

Steroid hormones

Teratocarcinomas
and embryonic stem cells

Transcription and translation

Virology

Yeast

Nucleic acid hybridisation

a practical approach

Edited by
B D Hames
Department of Biochemistry,
University of Leeds, Leeds, England
S J Higgins
Department of Biochemistry,
University of Leeds, Leeds, England

IRL PRESS
Oxford·Washington DC

IRL Press
Eynsham
Oxford
England

© IRL Press Limited 1985
First Published 1985
Reprinted 1986, 1987 (twice), 1988, 1990

British Library Cataloguing in Publication Data

Nucleic acid hybridisation : a practical approach. — (Practical approach series)
  1. Nucleic acid hybridisation
  I. Hames, B.D. II. Higgins, S.J. III. Series
  547.7'9   QP620

ISBN 0-947946-61-6 (Hardbound)
ISBN 0-947946-23-3 (Softbound)

Cover illustration. On the left is an autoradiograph of λgt11 recombinants containing *Dictyostelium discoideum* genomic DNA inserts, screened on nitrocellulose filters with a nick-translated 4.1 kb repetitive genomic fragment by the method of Benton and Davis (see Chapter 5). On the right is an autoradiograph of part of *Drosophila melanogaster* polytene chromosome 2R (stained with Giemsa) showing multiple sites of *in situ* hybridisation by the mobile element pDmI 137 [Dawid *et al.* (1981) Cell **25**, 399]; magnification 1300 x. The photographs were kindly supplied by Ms. P. Jagger and Mr. D.P. Ramji (Department of Biochemistry, University of Leeds, UK) and Dr. M.L. Pardue (M.I.T., Cambridge, USA), respectively.

Printed by Information Press Ltd, Oxford, England.

Preface

Nucleic acid hybridisation, the formation of a duplex between two complementary nucleotide sequences, is the basis for a range of techniques now in widespread use in modern biology. Hybridisation methodology extends from the quantitation of specific nucleic acid sequences, through their isolation and structural analysis involving recombinant DNA techniques, to investigations of their intracellular localisation, synthesis and regulation. The applications range from fundamental research in molecular biology to clinical diagnosis of human genetic disease. Our primary aim in this book has been to provide detailed practical protocols for the major hybridisation procedures. However, this alone is not sufficient. Optimal use of hybridisation requires a clear understanding of the principles and essential theoretical background so that the basic techniques can be modified to suit the particular purposes and conditions of the experiment. Therefore theory, rationale and practical advice are interwoven throughout this volume. We believe that the end result is both a useful guide for the novice and a valuable addition to the laboratories of more experienced researchers. Our thanks are due to the authors not only for the quality of their contributions but also for their patience and understanding during the numerous additions and modifications which we felt were necessary for completeness and to achieve a basic uniformity of approach and style.

David Hames and Steve Higgins

Contributors

M.L.M. Anderson
Wolfson Laboratory for Molecular Pathology, The Beatson Institute for Cancer Research, Garscube Estate, Switchback Road, Bearsden, Glasgow G61 1BD, UK

J.E. Arrand
Mammalian Cell DNA Repair Research Group, Department of Zoology, University of Cambridge, Downing Street, Cambridge CB2 3EJ, UK

R.J. Britten
Division of Biology, California Institute of Technology, Pasadena, CA 91125, USA

E.H. Davidson
Division of Biology, California Institute of Technology, Pasadena, CA 91125, USA

P.J. Mason
Imperial Cancer Research Fund, Burtonhole Lane, Mill Hill, London NW7 1AD, UK

S.J. Minter
Department of Applied Molecular Biology and Biochemistry, UMIST, P.O. Box 88, Manchester M60 1QD, UK

P. Oudet
Laboratoire de Génétique Moléculaire des Eucaryotes du CNRS, Unité 184 de Biologie Moléculaire et de Génie Génétique de l'INSERM, Faculté de Médecine, 11 Rue Humann, 67085 Strasbourg Cedex, France

M.L. Pardue
Department of Biology, Massachusetts Institute of Technology, 77 Massachusetts Avenue, Cambridge, MA 02139, USA

R.J. Roberts
Cold Spring Harbor Laboratory, P.O. Box 100, Cold Spring Harbor, NY 11724, USA

C. Schatz
Laboratoire de Génétique Moléculaire des Eucaryotes du CNRS, Unité 184 de Biologie Moléculaire et de Génie Génétique de l'INSERM, Faculté de Médecine, 11 Rue Humann, 67085 Strasbourg Cedex, France

P.G. Sealey
MRC Mammalian Genome Unit, Department of Zoology, University of Edinburgh, West Mains Road, Edinburgh EH9 3JT, UK

E.M. Southern
MRC Mammalian Genome Unit, Department of Zoology, University of Edinburgh, West Mains Road, Edinburgh EH9 3JT, UK

J.G. Williams
Imperial Cancer Research Fund, Burtonhole Lane, Mill Hill, London NW7 1AD, UK

B.D. Young
Department of Medical Oncology, Imperial Cancer Research Fund, St. Bartholomew's Hospital, West Smithfield, London EC1A 7BE, UK

Contents

APPENDICES

Abbreviations

AMV	avian myeloblastosis virus
ANLL	acute non-lymphocytic leukaemia
APT	O-aminophenylthioether
bp	base pairs
BSA	bovine serum albumin
Ci	Curie
C_ot	product of original concentration of nucleic acid and time
cDNA	complementary DNA
CLL	chronic lymphocytic leukaemia
CML	chronic myeloid leukaemia
c.p.m.	counts per minute
DBM	diazobenzyloxymethyl
DEAE	diethylaminoethyl
DEPC	diethylpyrocarbonate
DMSO	dimethyl sulphoxide
d.p.m.	disintegrations per minute
DTE	dithioerythritol
DTT	dithiothreitol
EDTA	ethylenediamine tetraacetic acid
EGTA	ethyleneglycobis(β-aminoethyl)ether tetraacetic acid
HAP	hydroxyapatite
Hepes	N-2-hydroxyethylpiperazine-N$'$-2-ethanesulphoric acid
h.p.l.c.	high performance liquid chromatography
IgV_λ	immunoglobulin λ light chain variable gene
kb	kilobases
mRNA	messenger RNA
nt	nucleotide(s)
PBS	phosphate-buffered saline
Pipes	piperazine-N,N$'$-bis-2-ethanesulphonic acid
R_ot	product of original concentration of RNA and time
rRNA	ribosomal RNA
SDS	sodium dodecylsulphate
SDS-PAGE	polyacrylamide gel electrophoresis in the presence of SDS
SSC	standard saline citrate
T_m	melting temperature
TCA	trichloroacetic acid
tRNA	transfer RNA

Introduction

EDWIN M. SOUTHERN

The discovery that the genetic message is written in a four letter code has had a profound influence on all areas of biology, and the ability to read that message by sequencing the DNA has given a technique of great power to biologists working in fields as diverse as medicine, agriculture and brewing. However, the genomes of most organisms contain a vast library of information and before analysis begins the particular DNA sequence of interest must first be isolated and characterised. It is here that molecular hybridisation gives us a great deal of help. The process which underlies all of the methods based on molecular hybridisation is the formation of the double helix from two complementary strands. This process was first described by Marmur and Doty in 1961. Two important features of the process were quickly established; the two sequences involved in duplex formation must have a degree of complementarity, and the stability of the duplex formed depends on the extent of the complementarity. These two features suggested ways in which the process could be used to analyse relationships between nucleic acid sequences and these were rapidly applied to a wide range of biological problems. Some methods, such as those developed by Nygaard and Hall and by Gillespie and Spiegelman, titrated the end-point of the reaction and this was used, for example, by Ritossa and Spiegelman to measure the number of ribosomal genes in *Drosophila melanogaster* by titrating DNA with ribosomal RNA. In combination with density gradient fractionation, the same method enabled Birnstiel to purify the genes for ribosomal RNA from *Xenopus laevis* long before the arrival of recombinant DNA techniques.

Other methods exploited the dependence of the rate of reaction on sequence complexity. When applied to DNA from higher organisms, it was found that a fraction reassociated at a rate much faster than expected from the high sequence complexity of the DNA. This fact led Britten and Kohne to the conclusion, surprising at the time and still not fully explained, that eukaryotic genomes contain a high proportion of repeated sequences. Rate measurements also formed the basis of the methods pioneered by the Carnegie group and by Bishop, for analysing the complexity of RNA populations in different cell types, which added important information to our understanding of the process of differentiation. For the student of evolution, the degree of relatedness between sequences, which could easily be determined by measuring the thermal stability of the duplexes, proved to be a powerful way of studying relationships between species and between genes.

In the late 1960s, Pardue and Gall, and Jones independently discovered a way of locating the position of specific sequences in the nucleus or chromosomes by carrying out the hybridisation reaction on cells fixed to microscope slides. In its original form, this method (*in situ* hybridisation) could be applied to only highly repeated sequences. Nevertheless, it produced some very important results, and in particular showed that some of the most highly repeated sequences in eukaryotic DNA are clustered in the heterochromatin. The method also paved the way for two important developments in the mid-1970s. At that time, recombinant DNA methods were at their beginning, and although the tremendous potential of the methods was widely recognised, this could not be realised fully without ways of detecting specific sequences in recombinant clones. Grunstein and Hogness provided this by applying molecular hybridisation directly to bacterial colonies, and shortly afterwards Benton and Davis devised a related method for phage plaques. These methods have had a tremendous influence on the pace of research. Gel electrophoresis of RNA and restriction fragments of DNA, introduced by Loening and by Danna and Nathans, respectively, was already well established as a powerful way of separating and analysing nucleic acids. Molecular hybridisation provided a way of detecting specific sequences among the bands (Southern and Northern blotting) and added much useful information to the analysis.

New analytical methods based on molecular hybridisation have played a major part in the rapid progress of biological research in recent years. Their use is expanding and other new methods continue to be developed. Nevertheless, common features underlie all the applications and it is pleasing to see that this volume, which places emphasis on the practicalities, also presents the fundamental aspects of the methods which will be important no matter what new developments arise. It is important to understand these principles, because no single set of conditions is optimal for all experiments. They guide the choice of the best experimental conditions and also help in solving problems that inevitably crop up from time to time. They form the starting point for the modification of existing techniques or the development of new ones. Besides understanding the theory, the practitioner will need a few recipes, and recipes that work well and consistently. Help will be needed in choosing between the various approaches that can be taken to the solution of a particular problem. This book contains a great deal of good advice and help with problems that occur even in the most advanced laboratories.

CHAPTER 1

Hybridisation Strategy

ROY J. BRITTEN and ERIC H. DAVIDSON

1. INTRODUCTION

At the present time, most hybridisation is done with radioactive probes and filter-bound nucleic acids and the underlying nature of the process is not of primary interest. Nevertheless, the basic principles and quantitative relationships need to be understood by researchers using hybridisation since, on occasion, they may be used to convert failure into success, or to obtain quantitative results. We have chosen to 'start at the beginning', define the terms required and list (without derivation) the most useful equations describing hybridisation. Our aim is entirely practical, so we have included a minimum of references and avoided the underlying physical chemistry and all abstruse issues. A considerably more detailed account of quantitative analysis of hybridisation, for those who require it, is given in Chapters 3 and 4.

2. DEFINITIONS

Single copy DNA is the set of DNA sequences that occur once per (haploid) genome, while *repeated DNA sequences* occur more than once. The physical separation of the two is not always easy, and often 'single copy DNA' describes a preparation which is contaminated with minor amounts of repeats.

Complexity describes the total length of different sequences present in a sample of nucleic acid. In the simplest case, where there is no sequence repetition, it equals the *genome size*, which, of course, is the complexity of one haploid complement of chromosomal DNA. The difference between the DNA sequences of sex chromosomes is generally not known and is ignored here since it can have only a small effect. Since most genomes are polymorphic in DNA sequence, the definition lacks absolute strictness. The amount of diverse sequence becomes a matter of definition and the judgement is based on the ability of sequences to hybridise with each other under the conditions of measurement. Obviously, if repeated sequences occur in the DNA of a genome, the same issues arise. Thus, in practice, complexity is the sum of the lengths of sequences so different that they do not pair with each other under the conditions of measurement. If a family of repeats is made up of many fairly similar copies of a given sequence, its complexity is just the length of the canonical sequence. Complexity has the same meaning for a population of RNA without regard to the prevalence (abundance) of different transcripts.

Reassociation is the joining together by typical base pairing of two fully-separated complementary sequences. Once reassociation has commenced, *zippering* describes the formation of successive base pairs. *Renaturation* has been used in both senses. *Hybridis-*

3

ation is now used to describe the formation of sequence-specific, base-paired duplexes from any combination of nucleic acid fragments (usually *in vitro*). A useful measure of the stability of a DNA duplex or an RNA-DNA hybrid is the *melting temperature* (T_m), that is, the temperature at which the strands are half dissociated or denatured.

$C_o t$ is the product of concentration of nucleic acid (mols of nucleotide per litre) and time (seconds). (The term $R_o t$ has also been used when RNA is the nucleic acid under consideration). Where a complex set of components with widely different rate constants is present, it is convenient to plot the fraction remaining unpaired against the logarithm of $C_o t$ and this has come to be known as a 'log $C_o t$ plot'.

Other symbols used in this chapter are defined in *Table 1*.

3. KINETICS OF REASSOCIATION

3.1 Practical Kinetic Equations

The reassociation of DNA sequences can be monitored by a number of methods including binding to hydroxyapatite, resistance to nuclease S1 digestion and optical hyperchromicity (see Chapter 3). Reassociation, as measured by these procedures, follows simple kinetic equations only approximately. Note the difference between DNA strands linked to duplex structures but not yet paired and DNA which is actually base paired.

Hydroxyapatite binding measures the fraction of DNA which is linked to structures containing duplexes and follows Equation 1, using the symbols listed in *Table 1*.

$$H = (1 + k\ C_o t)^{-1} \qquad\qquad \text{Equation 1}$$

Nuclease S1, which is single strand-specific, can be used to measure the fraction of DNA that has not yet reassociated (single-stranded). In the case of reassociation of a single, purified, short *restriction fragment*, bimolecular collisions control the rate of reassociation and lead to complete zippering of reassociating strands. Thus, second order kinetics following Equation 1 are expected and observed. Since each reassociated

Table I. Symbol Definitions

A	Fraction of total RNA that is accounted for by the species represented by the probe.
C_o	Original concentration of nucleotides (mol. litre^{-1}).
D	Reduction in melting temperature.
G	Complexity or genome size (number of nucleotides).
H	Fraction of DNA not bound to hydroxyapatite.
k	Observed rate constant (M^{-1}. sec^{-1}). The same symbol is used for a variety of conditions, methods of assay and kinetic functions.
L	Length of duplex.
L_1, L_2	Average length of DNA fragments.
P	Fraction of initially added probe that is hybridised in a probe-excess titration reaction.
R	Rate constant (M^{-1}. sec^{-1}). Independent of complexity. However, it is dependent on ionic strength, temperature, length, etc. A typical value is 1 000 000 for solution reassociation of fragments 500 nucleotides long in 0.18 M Na$^+$ at optimum temperature assayed by hydroxyapatite.
S	Fraction of nucleotides remaining unpaired.
t	Time in seconds.
X	Total mass of RNA added to a titration reaction.

restriction fragment is fully base-paired, hydroxyapatite and nuclease S1 give identical values for the reassociation and so S, the fraction of DNA resistant to nuclease S1, may be substituted for H in Equation 1 in this special case. However, in a typical reassociation reaction involving *randomly-sheared* DNA, the method of assay must be taken into account for the following reason. For a population of randomly-sheared DNA molecules having the usual wide range of fragment lengths, the *early* events of reassociation lead to duplexes that are paired over approximately 56% of their length, according to computer modelling (1). Such partially-reassociated molecules, however, will be regarded as fully-associated by hydroxyapatite whilst nuclease S1 will distinguish between the resistant base-paired portions and susceptible single-stranded regions. Thus the *early* stages of reassociation will follow Equation 1 if monitored by hydroxyapatite but the kinetics of reassociation will follow Equation 2 if monitored using nuclease S1.

$$S = (1 + k \, C_0 t)^{-0.44} \qquad \qquad \text{Equation 2}$$

Typically the value of k in Equations 1 and 2 is the same. Measurement of reassociation by *optical hyperchromicity* will also approximately follow Equation 2 with the same rate constant.

At *later* stages in the reassociation of randomly-sheared DNA, the single-stranded tails of one partially-reassociated duplex may become base-paired with the single-stranded regions of other partial duplexes. The result is hyperpolymer formation (2). The rate at which each duplex completes its reassociation (expressed per remaining single-stranded nucleotide) is then reduced as a result of steric hindrance (1,3,4). This moderately complex process is described approximately by simple equations. Assayed using hydroxyapatite, the kinetics follow Equation 1, although k is somewhat less than that for restriction fragments (described above). Assayed using nuclease S1, the kinetics of reassociation follow Equation 2.

3.2 Example of Reassociation

In our laboratory we still occasionally use the optical hyperchromicity method which is rapid and convenient for the determination of rates of reassociation under non-standard conditions. As an illustrative example, a *Hae*III digest of phage lambda DNA (48.5 kb 'complexity') was reassociated in 6 mM $MgCl_2$, 6 mM NaCl, 6 mM Tris-HCl, pH 7.5. The concentration was 24 μg/ml and the incubation temperature (80°C) was 10°C below the T_m. The optical assay showed 50% reassociation at 360 sec. The final drop in optical density was to 26% below the optical density of fully-denatured DNA (0.620). Equation 1 is applied since the DNA had been digested with a restriction enzyme and so no partially-paired duplexes should have been formed. The result obtained is that $k = 32$. The expected value of k for standard conditions can be calculated using Equation 3.

$$k = \frac{R}{G} \qquad \qquad \text{Equation 3}$$

where R is the rate constant ($M^{-1}. sec^{-1}$) irrespective of complexity and G is the complexity or genome size (number of nucleotides). Although R is independent of complexity, it is dependent on ionic strength, temperature, length, etc. A typical value is

10^6 M^{-1}. sec^{-1} for reassociation under standard conditions, that is, fragments 500 nucleotides long reassociating in solution in the presence of 0.18 M Na^+ at the optimum temperature and assayed by hydroxyapatite binding. Therefore, for the phage lambda DNA restriction fragment example being considered, the value of k for these standard conditions is given by:

$$k = \frac{10^6}{4.85 \times 10^4} = 21$$

The larger observed rate constant compared with that expected suggests that 6 mM Mg^{2+} ions accelerate hybridisation by a small factor (32:21). In fact, since the temperature of hybridisation was above optimal, the acceleration is probably more than a factor of two compared with standard conditions.

3.3 Pseudo First Order Kinetics

In certain circumstances, one of the complementary strands in excess does not reassociate and thus remains at constant concentration to drive the reassociation. This can be achieved with strand-separated fragments or approximated by immobilising the DNA, for example on a filter. Under these conditions, the kinetics for reassociation of the minority DNA fragments to the driver follows Equation 4. This is usually termed pseudo-first order kinetics.

$$S = e^{-kC_o t} \hspace{4cm} \text{Equation 4}$$

The symbols are as defined for Equations 1 and 2.

4. FACTORS AFFECTING REASSOCIATION

A number of factors can affect the *rate* of association and/or the *stability* of the duplexes formed.

4.1 Rate of Reassociation

4.1.1 *Temperature*

For a typical DNA association reaction, the graph relating reassociation rate to temperature shows a broad curve with a maximum rate at about 25°C below the melting temperature (T_m) of the duplexes (3,5,6). This is the optimum temperature for reassociation. Depending on the concentration of salt present (see Section 4.1.2), reassociation may effectively cease at temperatures well below the optimum. For example, the rate at which phage λ DNA reassociates in the presence of Mg^{2+} ions is reduced by at least two orders of magnitude when the reaction is carried out at a temperature 55°C below T_m.

4.1.2 *Concentration of Salt*

The concentration of salt affects the rate of reassociation very markedly (5). Below 0.1 M NaCl, a 2-fold increase in concentration increases the rate by 5- to 10-fold or more. The rate continues to rise with salt concentration but becomes constant when the concentration exceeds 1.2 M NaCl, that is, about seven times the concentration present under standard conditions (0.18 M Na^+). Thus the value of R in Equation 3 becomes approximately 8×10^6 M^{-1}. sec^{-1}. Divalent cations, which are often present as impurities in solutions, have a much more pronounced effect than monovalent ions at low concentrations and judicious use of EDTA as a chelator may be required, unless the monovalent ion concentrations are high.

4.1.3 Base Mismatch

The precision with which base-pairing occurs between reassociating DNA strands also affects the rate of reassociation (6). Thus, for each 10% mismatch the rate is reduced by a factor of about 2 when the reassociaton is carried out at the temperature that is optimal for the mismatched sequences, that is, 25°C below their T_m (Section 4.1.1).

4.1.4 Fragment Length

DNA-DNA reassociation. The effect of length on the rate of reassociation is well known for most cases of reassociation in solution (length effects are not well understood for filter hybridisation). Under most practical conditions, the rate of zippering is very fast compared with the rate of formation of the initial short specifically base-paired region ('nucleation'). (At high salt, low complexity, and high DNA concentration, a zero order zippering reaction can be observed, but such extreme conditions are unimportant in practice and are ignored throughout this chapter. All comments refer to nucleation-limited reactions.) When both complementary fragments are of the same size, the rate of reassociation rises as the square root of the length as shown by Equation 5 where L_1 and L_2 are the average lengths of the DNA fragments. This equation applies to reassociation under standard conditions and measured by any method of assay.

$$k\,(L_1) = k\,(L_2)\,x\left(\frac{L_1}{L_2}\right)^{\frac{1}{2}} \qquad \text{Equation 5}$$

This result is apparently due to the combination of an increased number of possibilities for nucleation per fragment balanced against increased steric hindrance or 'shadowing' for larger fragments, that is, the 'excluded volume effect' (3). When the two complements are different in length, the situation is more complicated. Here the effect of length on rate depends on which fragment is in excess. When the driver fragment is short and the tracer fragment is long, the result is as expected; the rate increases in proportion to the tracer fragment length (7). In other words, the nucleation rate appears to remain constant while the amount of the tracer linked to duplex rises with the fragment length. This effect is seen when using the hydroxyapatite assay but not with the nuclease S1 assay for the obvious reason that the long unpaired single-stranded tails are digested. It should also be true where the reassociation involves pairing with driver DNA bound to a filter. However, this case has not, to our knowledge, been tested. When the tracer fragment is short and the driver fragment is long, there is an unexpected reduction in rate (7). These situations are not discussed in detail here. Readers who have need for a quantitative examination of rates in reactions involving reactants of different lengths are referred to the original papers (1,4,7 − 9). However, in practice, these variations have little effect on the use of reassociation provided they are approximately taken into account in calculating the time necessary to approach completion.

RNA-DNA hybridisation. The hybridisation of RNA to a complementary DNA strand follows somewhat different rules, presumably as a result of the greater amount of secondary structure in RNA due to random coil self-interactions. When RNA is in excess and is hybridised with DNA in trace amounts under standard conditions (0.18 M NaCl), the rate of reassociation observed is almost the same as with DNA-DNA reassociation (8). However, as the salt concentration is raised, the rate of hybridisation driven by excess RNA does not rise as fast as the rate of DNA-DNA reassociation (10).

An unexpected result occurs when RNA-DNA hybridisation reactions are driven with excess DNA. The observed rate is 4- to 5-fold slower than that expected from DNA-DNA reassociation (4,9,10). This phenomenon, and the reduction in rate that occurs with excess long DNA fragments driving short tracer fragments (see above), are not understood. However, as for DNA-DNA reassociation, these variations in rate have little effect on the use of hybridisation in practice, so long as they are approximately taken into account in considering the extent of reaction necessary to reach completion.

4.1.5 *Complexity*

For simple genomes (such as those of phage, viruses, bacteria) where there is no significant sequence repetition, the rate of DNA reassociation follows Equations 1 or 2 and is inversely proportional to the complexity (or genome size). For complex genomes (such as those of eukaryotes), where considerable sequence repetition occurs, reassociation kinetics are much more complex. The single copy DNA (non-repetitive DNA) of these complex genomes reassociates at a rate inversely proportional to the genomic size. If the single copy DNA is purified by removing the repeated sequences, it then reassociates at a rate which is inversely proportional to the complexity of that single copy component. This has practical consequences in the quantitative use of reassociation kinetics and is discussed further in Section 5.

4.1.6 *Special Conditions*

There are special conditions where the rate of reassociation is highly increased. Kohne *et al.* (11) using a two-phase system, probably dependent on adsorption at a phenol-water interface, have increased the rate of reassociation by a factor of more than 10 000 for small quantities of DNA. In our experience the rates and extent of reassociation are hard to control with this method, but for certain special purposes it is very powerful. This method cannot easily be applied to reactions involving filter-bound nucleic acids. However, other factors can be used to increase the rate of filter hybridisation reactions. The inclusion of inert high molecular weight polymers such as dextran sulphate, which effectively concentrates the nucleic acid probe in solution (12), and high concentrations of salt (3) are two examples. The useful range of acceleration available with each of these may be a factor of 10.

4.2 **Stability of Duplexes**

4.2.1 *Base-pairing Mismatch*

Under the conditions of reassociation, sequences which are not perfectly complementary may form duplexes. Since mismatched sequences are less stable than perfectly base-paired species, the conditions of incubation or later steps of washing or melting of duplexes can be used to control which strand pairs are stabilised and hence assayed. The best estimate is that a 1% mismatch reduces the thermal stability as measured by the duplex melting temperature (T_m), by 1°C (5,13).

4.2.2 *Fragment Length*

The length of the base-paired region of a duplex affects the thermal stability of the duplex. The magnitude of this effect can be estimated from Equation 6.

$$D = \frac{500}{L} \qquad \text{Equation 6}$$

where D is the reduction in T_m (°C) and L is the length of the base-paired duplex.

4.2.3 *Concentration of Salt*

The concentration of salt has a profound effect upon the T_m of a duplex. The T_m changes about 16°C for each factor of 10 in salt concentration in the lower concentration range (0.01 − 0.1 M), but the effect is reduced at high concentrations (1 M).

Divalent cations have a much greater effect on the T_m than do monovalent cations. Small amounts of contaminating divalent cations present as impurities in solutions can be critical in attempts to utilise low salt concentrations to distinguish duplexes of different stabilities.

4.2.4 *Base Composition*

Since in normal salt solutions GC base pairs are more stable than AT base pairs, the T_m of a particular duplex is related to its GC content according to Equation 7.

$$T_m = 0.41 \ (\% \ GC) + 69.3 \qquad \text{Equation 7}$$

However, in certain chaotropic solvents, the effect of base composition on T_m can be eliminated, as with 2.4 M tetraethyl ammonium chloride described in Section 4.2.5.

4.2.5 *Criterion*

The word *criterion* describes the lower limit of fidelity of base pairing and the length of duplex established by the incubation conditions. In certain conditions such as denaturation of DNA in 2.4 M tetraethyl ammonium chloride, which nearly eliminates the range in T_m due to base composition (Section 4.2.4), the criterion can be very well-defined and it is possible to denature duplexes with a T_m as little as 1°C below that of other duplexes which survive unaffected. However, criterion is usually much less well-defined. In typical salt conditions, the range of melting temperature due to differences in base composition of eukaryotic DNA is as much as 15°C. Other problems also arise. For example, when filter-bound duplexes are washed in low salt buffer below the criterion temperature, hybrid structures which should be denatured are only very slowly removed. In such cases, it is common to observe an uncertainty of more than 20°C in the effective criterion.

5. QUANTITATIVE USE OF KINETICS OF REASSOCIATION

5.1 Conditions of Analysis

As mentioned above (Section 4.1.5), the rate of reassociation is determined by the complexity of a DNA sample. Thus, the rate of reassociation can be used to estimate the genome size of an organism. For organisms with simple genomes lacking repetitive DNA sequences this is straightforward. For complex eukaryotic genomes the majority of the complexity is accounted for by the single copy DNA so the conditions of reassociation must be chosen to allow nearly complete reassociation of that component (5,14). The complexity is calculated by reference to a standard DNA from an organism of known complexity, such as a bacterial DNA. To guard against differences in the reaction conditions (salt concentration, viscosity, etc.) for the DNAs that are being compared, it is essential that the standard is included in the reaction of the sample DNA, that is, it must be used as an *internal* standard. This is easily carried out by using radio-actively-labelled standard DNA. The standard DNA should have approximately the same fragment size as the sample DNA and its concentration should be adjusted so that it reacts over a similar time period as the sample. Procedures for labelling DNA are described in Chapter 2.

If repeated sequences are present, then least squares methods are required to analyse the reassociation data so that the different components reacting in the reassociation process can be separated and the corresponding rates of reaction accurately determined (see Chapter 3 and ref. 15). The hydroxyapatite assay is normally used for this type of experiment. The nuclease S1 assay is not preferred because of the slow termination of components due to the kinetic form (Equation 3). Pseudo-first order kinetics in solution would be ideal for such analyses but in most cases this is impractical due to the absence of appropriate single-stranded driver sequences. The rates of reactions involving filter-bound nucleic acid and single-stranded probe in solution have not been adequately standardised for this purpose.

5.2 Reassociation of Single Copy DNA

When total unfractionated eukaryotic DNA is utilised, the rate of reassociation of the single copy component is determined by the (haploid) amount of DNA per cell, that is, the total quantity of DNA which contains one single copy of each of the most slowly reassociating sequences. If the DNA is fractionated to prepare the single copy DNA and then studied in a self-driven reassociation reaction, the rate is determined by the amount of single copy DNA per cell (complexity). The fractionation can be achieved by reassociation to a C_0t value at which repetitive sequences have formed duplexes and the single copy DNA is still single-stranded, followed by hydroxyapatite chromatography to separate the single copy DNA (Chapter 3, Section 7.2). Perhaps the most accurate procedure is to prepare labelled single copy DNA and reassociate this with a large excess of total unlabelled DNA driver. In this case the total genome size determines the rate of reassociation of the single copy tracer with the total DNA driver.

5.3 Reassociation of Repetitive DNA

With unfractionated DNA the rate of reassociation of the individual repetitive components is determined by their repetition frequency (the number of copies per genome).

Since the reassociation of each component approximately follows Equation 1 (or 2), the ability to resolve different repetitive components is severely limited. The results are usually plotted in the standard fashion against log C_0t. Each component requires a factor of about 100 in C_0t to proceed from 10% to 90% reassociation. Any components for which reassociation appears to proceed faster are artifacts (though published curves have often been drawn through such meaningless fluctuations).

It may be possible to determine the complexity of the various repetitive components in total DNA in the same way as for single copy DNA, that is, by separating them individually using the method of reassociation to a pre-determined C_0t value followed by hydroxyapatite chromatography (see above). However, this procedure is severely limited by the inability to separate families whose frequencies do not differ by a factor of at least 10. Even this requires multiple steps of reassociation and fractionation. It is far more satisfactory to clone a representative sample of the repeats and study the individual families, as described below.

While single copy DNA does not show a large amount of sequence divergence (16), repetitive DNA often does. Depending on the incubation conditions, these poorly related sequences can reassociate with each other to form mismatched duplexes. As described in Section 4.1.3, 10% base mismatch reduces the rate of reassociation by a factor of two and this introduces an element of uncertainty into the analysis of repetitive DNA by kinetic methods.

Measurement of the kinetics of reassociation determines, with moderate accuracy, the total number of members of a family of repeated sequences (17,18). For this purpose a cloned probe from the family is reassociated with total sheared DNA to a C_0t value adequate to ensure complete reassociation and samples are assayed by hydroxyapatite. Other methods for determining the frequency of a repeated sequence include screening libraries with the cloned probe or quantitative estimation of the extent of reaction of the probe with filter-bound DNA (19−21). Screening, of course, measures only dispersed repeats and thus underestimates clustered copies. Measurement by the saturation of bound DNA on filters (DNA dot blots, see Chapter 4) holds promise for the future if the accuracy can be improved. By calibration, appropriate adjustment of conditions and driving the reaction with a sufficient excess of the probe over the bound DNA, one might obtain a good estimate of the total number of homologous sequences at a particular criterion.

6. QUANTITATIVE HYBRIDISATION OF RNA

Quantitative hybridisation procedures are widely used for determination of the abundance and sequence complexity of RNAs. We restrict our discussion to two problems, the measurement of RNA prevalence by titration hybridisations in solution, and the measurement of complexities of RNA populations.

6.1 Measurement of RNA Abundance

If a solution hybridisation reaction is carried to kinetic completion and one complementary strand is present in significant excess, then essentially all of the minority complementary strand will be paired. The result is not guaranteed for filter reactions since some bound sequences are not fully accessible and completion of the reactions is difficult to obtain. Nevertheless, dot blots on filters (see Chapter 4) are useful for quantitative determination with appropriate calibration (22).

The most exact and generally useful method for determining the prevalence of a given RNA species is titration in solution, using a cloned, labelled, single-stranded probe present in excess. This procedure differs fundamentally from the cDNA hybridisation method (23), in that the prevalence determination is not derived from the rate of the reaction but from the final extent of reaction. The RNA preparation that includes the molecular species of interest is mixed at various mass ratios with the excess tracer, and reacted to completion in solution. In practice, reaction conditions exceeding by 10-fold the half reaction point are easy to obtain, since the reactions are pseudo-first order (Equation 4) with respect to the excess cloned probe, which is always of low complexity. Thus, all the samples have essentially the same rapid reaction kinetics. This removes most of the uncertainties and difficulties associated with prevalence determination by kinetic methods, in which it is often difficult to be certain that the hybridisation reactions are actually completed (10), a problem that is particularly acute for rare RNA species that require very extensive hybridisation for their reactions (high C_ot; Equation 1). The amount of hybridisation obtained in each sample is proportional to the RNA/probe ratio. Therefore the abundance of the specific RNA in the RNA sample can be calculated directly from the extent of probe hybridisation as shown in Equation 8 (ref. 24).

$$P = AX \hspace{3cm} \text{Equation 8}$$

where P is the fraction of the probe added initially (in a probe-excess titration reaction), A is the fraction of the total RNA that is complementary to the probe and X is the total mass of RNA added to the titration reaction. Hybridisation is preferably measured by a nuclease method so that the number of nucleotides actually paired ($[1-S]$; Equation 2) is determined but if appropriate length corrections are made, hydroxyapatite values ($[1-H]$; Equation 1) can also be used to establish the value of P (ref. 25).

The relationship shown in Equation 8 is valid for the usual situation in which there are RNA transcripts representing only one genomic DNA strand, that is, that are complementary to the probe. In certain circumstances this is not the case, for instance when the probe is a typical interspersed repetitive sequence. Scheller *et al.* (24) and Costantini *et al.* (26) showed that such repeat sequences are usually represented by transcripts complementary to both strands of the canonical genomic repeat element. This is due to contributions from many different repeat sequence elements in both directions with respect to the transcription unit of which they are a part, and not to symmetrical transcription of any given repeat sequence element (27). In situations where both strands are represented in the RNA, the RNA complementary to the probe can also hybridise to other transcripts of the same sense as the probe, and an equation more complex than Equation 8 describes the reaction (24). However, the initial rate is still given by Equation 8 providing a simple method in which a non-linear least squares calculation with a computer is not required.

The limiting factor in application of the titration method is availability of the appropriate single-stranded DNA probe. A variety of methods can be used for radioactive labelling (Chapter 2) but the probes must be uniformly labelled in internal positions if a nuclease assay procedure is to be utilised. Almost any cloned DNA fragment can be strand-separated and then labelled *in vitro* with radioactive iodine (e.g., ref. 25), but in our experience this process is often time-consuming and difficult. At present the preferred methods are preparation of a labelled single-stranded DNA probe using an M13 vector (28), or of a labelled single-stranded RNA probe, using a vector based on the SP6 phage promoter (29,30).

In some situations the RNA is available in such quantity that excess cDNA can be prepared and used as the probe (e.g., ref. 31). RNA-RNA hybrids are somewhat easier to assay than RNA-DNA hybrids because of the relative resistance of RNA-RNA duplexes to RNase, compared with the sensitivity of RNA-DNA hybrids to nuclease S1 unless the concentrations of nuclease S1 and salt and the temperature are all carefully adjusted.

The solution titration method is recommended for the calibration of other methods that provide only relative measurements such as RNA gel blots, RNA dot blots, cDNA clone, colony hybridisations, etc. (e.g., refs. 32,33). Thus if the specific activity of the cloned tracer, and the mass of total RNA per cell are known, the absolute number of molecules of the hybridising RNA species per cell is estimated directly by measurement of parameter A in Equation 8.

6.2 Measurement of RNA Complexity

The original method for determination of the complexity of a population of RNA (the single copy saturation procedure; ref. 34) has not been fundamentally improved. The first step is the preparation of a short (200 – 500 nucleotide) labelled DNA fragment tracer containing all of the single copy DNA sequences of the genome and from which most of the repeated sequences have been removed, for example by using the method of reassociation to a pre-determined C_0t value followed by hydroxyapatite chromatography (see Section 5.2). This tracer is then hybridised with a large *sequence* excess of the RNA. The complexity of the RNA population is obtained directly from the saturation value, that is, the fraction of the single copy DNA tracer driven into hybrid at the completion of the reaction. This is conveniently assayed by hydroxyapatite chromatography, though nuclease S1 assays can also be utilised. In such measurements, each component will hybridise according to pseudo-first order kinetics (Equation 4) with the rate determined by its sequence concentration. The most complex fraction of the RNA population will therefore react last since this component contains the largest number of diverse sequences, each present at low concentration. Thus, incubation to high RNA C_0t is usually necessary. It is important to be certain that the reactions have reached completion as a result of hybridisation of all available complementary tracer molecules, rather than for artifactual reasons such as degradation of the RNA driver or of the DNA tracer. This procedure has been used to provide accurate measurements of the complexity of mRNA (e.g., refs. 35 – 38) and nuclear RNA populations (38,39) for many different cell types and organisms. The sensitivity of measurements such as these extends down to about one molecule of each RNA species per 10 – 100 cells, though in most cases (except mammalian brain) the lowest prevalence class actually observed is of the order of one to a few molecules per cell.

The strength of the single copy tracer saturation method is that the rate of reaction itself is not important, only the total amount of single copy DNA hybridised at completion. Thus, for measurement of the sequence diversity or complexity of an RNA population, the single copy saturation procedure is clearly more reliable than methods that depend directly on hybridisation *kinetics*, for example, hybridisation of trace amounts of cDNA back to its own template RNA population which is subject to large errors in estimation of the complexity of rare sequence components. With kinetic methods any factors that retard the reactions [e.g., a decrease in sequence concentration due to partial transcript degradation, inadequate mixing (10), changes in viscosity or length disparity (40)] will have severe consequences, by causing miscalculation of the kinetic components from which the complexity is calculated, and some unfortunate examples exist in the literature. The more prevalent, rapidly reacting components are least affected by any of those problematical factors. On the other hand, single copy tracer hybridisation in RNA excess is very insensitive for the analysis of prevalent, relatively low complexity transcript classes since they often account for only a small amount of hybridisation that is difficult to detect, compared with the far more diverse high complexity components. Instead, cDNA kinetics analysis is the method of choice for determining the quantities of major or average prevalence components within an RNA population. Thus, for most purposes the two methods are complementary; cDNA kinetics should be used to obtain the prevalence structure of the RNA population and single copy saturation hybridisation to obtain its true sequence diversity, or complexity. However, it should be noted that the latter objective can also be achieved using a cDNA tracer, provided that the more prevalent components are first removed and that exceptional precautions are taken with regard to mixing, salt concentrations, and other experimental variables (10).

7. ACKNOWLEDGEMENTS

Research from this laboratory was supported by grants from NSF, NICHHD and NIGMS.

8. REFERENCES

1. Britten,R.J. and Davidson,E.H. (1976) *Proc. Natl. Acad. Sci. USA,* **73**, 415.
2. Britten,R.J. and Waring,M.J. (1965) *Carnegie Inst. Wash. Yearbook,* **64**, 316.
3. Wetmur,J.G. and Davidson,N. (1968) *J. Mol. Biol.,* **31**, 349.
4. Smith,M.J., Britten,R.J. and Davidson,E.H. (1975) *Proc. Natl. Acad. Sci. USA,* **72**, 4805.
5. Britten,R.J., Graham,D.E. and Neufeld,B.R. (1974) in *Methods in Enzymology,* Vol. **29E**, Grossman,L. and Moldave,K. (eds.), Academic Press Inc., New York/London.
6. Bonner,T.I., Brenner,D.J., Neufeld,B.R. and Britten,R.J. (1973) *J. Mol. Biol.,* **81**, 123.
7. Chamberlin,M.E., Galau,G.A., Britten,R.J. and Davidson,E.H. (1978) *Nucleic Acids Res.,* **5**, 2073.
8. Galau,G.A., Britten,R.J. and Davidson,E.H. (1977) *Proc. Natl. Acad. Sci. USA,* **74**, 1020.
9. Galau,G.A., Smith,M.J., Britten,R.J. and Davidson,E.H. (1977) *Proc. Natl. Acad. Sci. USA,* **74**, 2306.
10. Van Ness,J. and Hahn,W.E. (1982) *Nucleic Acids Res.,* **10**, 8061.
11. Kohne,D.E., Levison,S.A. and Byers,M.J. (1977) *Biochemistry (Wash.),* **16**, 5329.
12. Chang,C.-T., Hain,T.C., Hutton,J.R. and Wetmur,J.G. (1974) *Biopolymers,* **13**, 1847.
13. Hutton,J.R. and Wetmur,J.G. (1973) *Biochem. Biophys. Res. Commun.,* **52**, 1148.
14. Britten,R.J. and Kohne,D.E. (1968) *Science (Wash.),* **161**, 529.
15. Pearson,W.R., Davidson,E.H. and Britten,R.J. (1977) *Nucleic Acids Res.,* **4**, 1727.
16. Britten,R.J., Cetta,A. and Davidson,E.H. (1978) *Cell,* **15**, 1175.
17. Klein,W.H., Thomas,T.L., Lai,C., Scheller,R.H., Britten,R.J. and Davidson,E.H. (1978) *Cell,* **14**, 889.
18. Moore,G.P., Pearson,W.R., Davidson,E.H. and Britten,R.J. (1981) *Chromosoma,* **84**, 19.
19. Thomas,T.L., Britten,R.J. and Davidson,E.H. (1982) *Dev. Biol.,* **94**, 230.

20. Anderson,D.M., Scheller,R.H., Posakony,J.W., McAllister,L.B., Trabert,S.W., Beall,C., Britten,R.J. and Davidson,E.H. (1981) *J. Mol. Biol.*, **145**, 5.
21. Dowsett,A.P. (1983) *Chromosoma*, **88**, 104.
22. Kafatos,F.C., Jones,C.W. and Efstratiadis,A. (1979) *Nucleic Acids Res.*, **7**, 1541.
23. Bishop,J.O. (1972) *Biochem. J.*, **126**, 171.
24. Scheller,R.H., Costantini,F.D., Kozlowski,M.R., Britten,R.J. and Davidson,E.H. (1978) *Cell*, **15**, 189.
25. Lev,Z., Thomas,T.L., Lee,A.S., Angerer,R.C., Britten,R.J. and Davidson,E.H. (1978) *Dev. Biol.*, **76**, 322.
26. Costantini,F.D., Scheller,R.H., Britten,R.J. and Davidson,E.H. (1978) *Cell*, **15**, 173.
27. Posakony,J.W., Scheller,R.H., Anderson,D.M., Britten,R.J. and Davidson,E.H. (1981) *J. Mol. Biol.*, **149**, 41.
28. Messing,J., Crea,R. and Seeburg,P.H. (1981) *Nucleic Acids Res.*, **9**, 309.
29. Kassavetes,G.A., Butler,E.T., Roulland,D. and Chamberlin,M.J. (1982) *J. Biol. Chem.*, **257**, 5779.
30. Butler,E.T. and Chamberlin,M.J. (1982) *J. Biol. Chem.*, **257**, 5772.
31. Wallace,R.B., Dube,S.K. and Bonner,J. (1977) *Science (Wash.)*, **198**, 1166.
32. Lasky,L.A., Lev,Z., Xin,J.-H., Britten,R.J. and Davidson,E.H. (1980) *Proc. Natl. Acad. Sci. USA*, **77**, 5317.
33. Xin,J.-H., Brandhorst,B.P., Britten,R.J. and Davidson,E.H. (1982) *Dev. Biol.*, **89**, 527.
34. Davidson,E.H. and Hough,B.R. (1971) *J. Mol. Biol.*, **56**, 491.
35. Galau,G.A., Britten,R.J. and Davidson,E.H. (1974) *Cell*, **2**, 9.
36. Galau,G.A., Klein,W.H., Davis,M.M., Britten,R.J. and Davidson,E.H. (1976) *Cell*, **7**, 487.
37. Hereford,L.M. and Rosbash,M. (1977) *Cell*, **10**, 453.
38. Bantle,J.A. and Hahn,W.E. (1979) *Cell*, **8**, 139.
39. Hough,B.R., Smith,M.J., Britten,R.J. and Davidson,E.H. (1975) *Cell*, **5**, 291.
40. Chamberlin,M.E., Galau,G.A., Britten,R.J. and Davidson,E.H. (1978) *Nucleic Acids Res.*, **5**, 2073.

CHAPTER 2

Preparation of Nucleic Acid Probes

JANET E. ARRAND

1. INTRODUCTION

The other chapters in this volume describe in detail the various types of hybridisation commonly in use; solution hybridisation, filter hybridisation and *in situ* hybridisation. The advantages and disadvantages of each system are fully explained, as are the situations in which each sort of hybridisation is used and the necessary probes. This chapter concentrates on the experimental protocols necessary to produce probes for each application. A chapter of this kind can never be fully comprehensive; new radiolabelling techniques are constantly being developed and the advent of non-radioactive detection systems adds a new dimension to the preparation, storage and use of nucleic acid probes. The aim has been to provide a compilation of proven techniques for the preparation of probes and advice on their use.

2. CHOICE OF NUCLEIC ACID PROBE FOR HYBRIDISATION

Any form of nucleic acid can be used as a probe for hybridisation provided that it can be suitably labelled. The choice of probe depends on three factors:

(i)   the hybridisation strategy;
(ii)   the availability or source of material for use as a probe;
(iii)   the degree to which it can be labelled.

Messenger RNA can be isolated from eukaryotic cells for use as a probe by techniques described in Section 3.1. The mRNA preparation can be enriched with respect to the mRNA species of interest either naturally (as is globin mRNA in red blood cells) or artificially by hybridisation to immobilised DNA. Although mRNA can be used as a probe for any type of hybridisation, there are limitations:

(i)   the mRNA may not be 100% pure. This will increase the background hybridisation and may make interpretation of the results more difficult.
(ii)   preparation of mRNA is time consuming; yields are low and the product is very easily degraded by exogenous and endogenous nucleases.
(iii)   the mRNA can be labelled by only one method, that is, using $[\gamma\text{-}^{32}\text{P}]$ATP and T4 polynucleotide kinase (Section 4.1.4).

In some instances, however, mRNA is the only available probe. Other RNAs (e.g., viral RNA, tRNA, rRNA) can be isolated by standard techniques and also used as probes. Most recently, *in vitro* RNA transcripts are now proving to be extremely effective hybridisation probes (see SP6 polymerase labelling, Section 4.2.2).

Once purified mRNA has been prepared, single-stranded cDNA hybridisation probes can be made using avian myeloblastosis virus (AMV) reverse transcriptase in the presence of labelled nucleoside triphosphates (see Section 3.2). Alternatively, double-stranded cDNA copies can be made and cloned into plasmids or bacteriophages (see companion volumes in this series on DNA cloning, refs. 1 and 2). This increases yield and decreases vulnerability to attack by nucleases so that probe stability ceases to be a problem. Repetitive sequences, which can cause high background problems, can be located in double-stranded DNA by restriction enzyme mapping, and can be excised from the probe (Section 3.4.2) before hybridisation. Furthermore, any labelling method can be used for cloned cDNA probes. As well as serving as probes for the encoded sequence, these are particularly useful in the isolation of their genomic equivalents. The limitation for the use of cDNA clones lies in the nature of the vector since this may contain sequences which cross-hybridise with other nucleic acids in the test sample. In this case, it may be necessary to remove the vector sequences by restriction enzyme digestion prior to labelling and hybridisation.

Any DNA (cloned or uncloned) can be used as a probe for hybridisation. The techniques of restriction enzyme mapping (3) and nuclease S1 mapping (4) can provide a detailed knowledge of the organisation of any particular piece of DNA and thus aid the choice of a particular region for use as a probe.

Recently, detailed knowledge of the sequence of many proteins has enabled the construction of short homologous nucleic acid sequences (5). These synthetic oligonucleotides have proved invaluable for use as hybridisation probes (6). The longer the sequence, the more specific the probe. These oligonucleotide probes can be labelled and used to probe cDNA and genomic libraries for the gene coding for the protein of interest.

3. PREPARATION OF NUCLEIC ACIDS FOR USE AS HYBRIDISATION PROBES

3.1 Preparation of Eukaryotic Messenger RNA

RNA must be handled with great care to avoid degradation. All glassware should be washed thoroughly, rinsed with distilled water containing 0.1% diethylpyrocarbonate (DEPC), covered with aluminium foil and autoclaved. DEPC destroys RNase activity and excess DEPC is converted to ethanol and CO_2 by autoclaving. Solutions should be prepared using high purity chemicals, made 0.1% in DEPC and autoclaved. SDS is also an excellent RNase inhibitor and should be included in solutions at a concentration of $0.1-1\%$ where possible. Other RNase inhibitors which are useful for the protection of mRNA during preparation are vanadyl ribonucleoside complex, which is effective at a concentration of 10 mM, and human placental ribonuclease inhibitor (effective at $50-100$ units/ml.).

There are several alternative methods for the isolation of RNA depending on the source of the material. A typical method for the preparation of total cellular RNA from tissue culture cells or whole tissue is described below. It can be used for fresh, frozen or freeze-dried tissue.

(i) Suspend the tissue in 20 volumes of solution A (6 M guanidinium chloride or guanidinium isothiocyanate: 2 M potassium acetate pH 5.0; 19:1 v/v).

(ii) Sonicate three times, each time for 15 sec to shear the DNA. The solution should become non-viscous.

(iii) Centrifuge at 14 000 *g* for 5 min at 4°C to remove any insoluble debris.

(iv) Add 1/2 volume of 95% ethanol. Leave at −20°C overnight.

(v) Centrifuge at 14 000 *g* for 15 min at −10°C to pellet the RNA.

(vi) Dissolve the pellet in 1/10th the original volume of solution A containing 25 mM EDTA. Add 1/2 volume of 95% ethanol and leave at −20°C for 2 h. A flocculent precipitate should form.

(vii) Centrifuge at 14 000 *g* for 15 min at −10°C.

(viii) Repeat steps (vi) and (vii). This selectively precipitates RNA and leaves DNA and protein in solution.

(ix) Dissolve the pellet in a small volume of 20 mM EDTA (1−2 ml) at 0°C. This step requires homogenisation with a tight-fitting Dounce homogeniser.

(x) Extract the solution with an equal volume of chloroform:butan-1-ol (4:1 v/v). Shake at room temperature for 5 min. Centrifuge at 14 000 *g* for 5 min to separate the phases. Re-extract the organic phase with 1−2 ml of 20 mM EDTA and pool the aqueous phases.

(xi) Add 1/5th volume of 4 M NaCl and 2 volumes of ethanol. Leave at −20°C overnight.

(xii) Store the RNA in the 70% ethanol mixture until it is required.

(xiii) To recover the RNA, centrifuge at 14 000 *g* for 15 min at −10°C. Air-dry the pellet and dissolve immediately in sterile water or an appropriate buffer.

Table 1. Purification of Poly(A)$^+$ mRNA by Chromatography on Oligo(dT)-cellulose.

Preparation of oligo(dT)-cellulose column

1. Plug the end of a 5 ml disposable plastic syringe with siliconised glass wool.
2. Suspend commercial, dry oligo(dT)-cellulose in TS buffer (10 mM Tris-HCl, pH 7.4, 0.1% SDS). Transfer sufficient of this to the syringe to give 1−2 ml bed volume.
3. Wash the column with several bed volumes of TS buffer. Add 0.1 M NaOH to the column and allow it to run into the matrix.
4. Stand the column for 30 min at room temperature.
5. Wash off the NaOH with several column volumes of NTS buffer (0.5 M NaCl, 10 mM Tris-HCl, pH 7.4, 0.1% SDS).
6. The column is now ready for use at room temperature or for storage at 4°C.

Fractionation of mRNA

1. Suspend the total RNA (prepared as in Section 3.1) in 5 ml of NTS buffer.
2. Apply this RNA solution to the oligo(dT)-cellulose column and collect the flow-through buffer [which will contain mainly poly(A)$^-$ RNAs].
3. Re-apply this to the column twice more.
4. Wash the column with NTS buffer until no RNA (detected by absorbance at 260 nm) can be detected in the flow-through buffer.
5. Elute the bound mRNA with 5 ml of TS buffer. Read the A_{260} and estimate the yield; it should be about 2−5% of the total RNA loaded.
6. Add NaCl to the RNA solution to a final concentration of 0.5 M (i.e., the RNA is now in NTS buffer).
7. Re-equilibrate the column with NTS buffer.
8. Carry out a second cycle of binding and elution on the oligo(dT)-cellulose column, that is, apply the RNA from step 6 and repeat steps 3−5.
9. Ethanol precipitate, store and collect the mRNA as in Section 3.1, steps (xi)−(xiii).

Most eukaryotic mRNA molecules have a 3′ poly(A) tract [poly(A)+ mRNAs] and can therefore be separated from other RNAs lacking poly(A) [poly(A)− mRNAs] by selective binding to and elution from oligo(dT)-cellulose or poly(U)-Sepharose. A protocol for oligo(dT)-cellulose chromatography is given in *Table 1*. Poly(A)− mRNAs, such as some histone mRNAs, do not bind to oligo(dT)-cellulose and so can be collected in the flow-through buffer together with non-messenger RNA species such as tRNA and rRNA. Fractionation of cells into nuclear and cytoplasmic elements prior to RNA purification may enrich for poly(A)− mRNAs (see ref. 7). The genomes of some RNA viruses,such as influenza virus,'are composed of poly(A)− RNA segments which act directly as mRNA upon infection. These are readily purified from the virion particles.

3.2 **Preparation of Single-stranded cDNA Probes**

Synthesis of cDNA is carried out on a poly(A)+ mRNA template using oligo(dT) hybridised to the poly(A) tail to prime complementary strand synthesis by the RNA-dependent DNA polymerase, AMV reverse transcriptase. A typical protocol is given in *Table 2*. In order to make a cDNA copy of poly(A)− mRNA, a poly(A) tail must first be added using *Escherichia coli* poly(A) polymerase in the following reaction.

(i)   Adjust the mRNA concentration so that the concentration of 3′ ends is 100 nM (i.e., 10 μl will contain 1 pmol of 3′ ends. If the concentration of 3′ ends is higher than this, the concentration of poly(A) polymerase (see below) must be increased proportionally.

Table 2. Synthesis of cDNA.

1.	Dry down 10 μl of a 5 mM solution of each of the four deoxyribonucleotides (dATP, dCTP, dGTP, dTTP) plus a small quantity (<1 μCi) [α-^{32}P]dATP (400 Ci/mmol) which is used to follow the extent of synthesis.

2.	Add:
	     20 μl 5 x RTase buffer (0.25 M KCl, 30 mM MgCl$_2$, 0.25 M Tris-HCl, pH 7.2)
	     10 μl 50 mM DTT
	     10 μl oligo(dT)$_{12-18}$ (10 A_{260} units/ml)
	     10 μl 500 U/ml placental RNase inhibitor (commercially available)
	     25−50 units of AMV reverse transcriptase [a]
	     1−2 μg mRNA
	     Water to a final volume of 100 μl

3.	Incubate at 42°C for 1 h.
4.	Stop the reaction by adding 1/10 volume of 0.1 M EDTA and 1/15th volume of 5 M NaOH.
5.	Incubate at 60°C for 30 min.
6.	Add 1/10th volume of 1 M Tris-HCl, pH 7.5, and enough 1 M HCl to neutralise the solution.
7.	Chromatograph through Sephadex G-50 (packed in a siliconised Pasteur pipette) in 5 mM NaCl, 5 mM Tris-HCl, pH 7.5. Collect 4-drop fractions (~0.2 ml). Follow the elution of the radioactive peaks using a hand monitor. The cDNA will be in the leading peak and unincorporated nucleotides will be in the trailing peak. Determine the radioactivity of the fractions in the collection tubes by Cerenkov counting.
8.	Recover the cDNA by ethanol precipitation as in Section 3.1, steps (xi) and (xiii).

[a]The units of reverse transcriptase quoted here are those defined by Dr. J.W.Beard of Life Sciences Inc. One unit is the amount of enzyme which incorporates 1 nmol of dTMP into an acid-insoluble product in 10 min at 37°C using poly(rA):oligo(dT)$_{12-18}$ as template. Take care since some suppliers use different units.

(ii)   Mix:

20 μl 5 x polyadenylation buffer (2.5 mg/ml BSA, 0.1 M MgCl$_2$, 0.25 M Tris-HCl, pH 7.9).

2.5 μl 0.1 M MnCl$_2$

2.5 μl 10 mM rATP (containing a small, defined amount of $[\alpha-^{32}P]$rATP ($\sim$5000 c.p.m.)

2 μl 5 M NaCl

10 μl mRNA solution from step (i)

50 μl water

10 μl poly(A) polymerase (diluted to 0.4 U/ml)

(iii)   Incubate for 10 min at 37°C. This should result in the synthesis of poly(A) tails up to 50 residues in length. (The absolute tail length may be estimated by measuring the incorporation of $[^{32}P]$rATP per pmol of 3' ends.)

(v)   Recover the polyadenylated mRNA by chromatography on G50 Sephadex in 50 mM NaCl, 1 mM EDTA, 10 mM Tris-HCl, pH 7.5, collecting 0.2 ml fractions. The poly(A)-tailed RNA elutes in the excluded volume and can be located by Cerenkov counting.

(vi)   Precipitate the RNA, recover and store it as in Section 3.1, steps (xi)−(xiii).

3.3 Isolation of Genomic DNA

High molecular weight, genomic DNA can be prepared from animal, plant and bacterial cells by essentially similar techniques. The steps involved are:

(i)   cell lysis;

(ii)   removal of proteins (and polysaccharides in the case of plant material);

(iii)   removal of RNA;

(iv)   recovery of pure DNA.

A method for the preparation of high molecular weight DNA from mammalian cells or disrupted tissue is given below. For the preparation of plant and bacterial DNA, the method of cell lysis may be different but other steps will be essentially the same. Whatever the source of the DNA, several problems arise because of its size. Firstly, the DNA solutions are very viscous making them difficult to pipette. This problem is overcome by using wide-bore pipettes. For disposable pipettes and micropipette tips, the ends can be cut off to provide a wider aperture. Secondly, the DNA is very sensitive to shear. Therefore, to avoid breakage of the large DNA molecules, never use a vortex mixer or a fast shaker during DNA preparation, and always pipette gently.

(i)   Harvest the tissue culture cells by low speed centrifugation (*or* gently disrupt small pieces of tissue by stirring at 4°C for 10 min in 4% EDTA, 0.15% Tris).

(ii)   Wash the cells with phosphate-buffered saline (8.0 g NaCl, 0.2 g KCl, 1.5 g NaH$_2$PO$_4$, 0.2 g K$_2$HPO$_4$ per litre of solution) by centrifugation. Freeze the cell pellets at −20°C.

(iii)   Thaw the cells into 20 volumes of 0.1 M EDTA, 10 mM Tris-HCl, pH 7.4. (This unusually high concentration of EDTA inhibits the action of endogenous nucleases.)

(iv)   Add SDS to a final concentration of 0.5% and mix gently.

(v)   Add Proteinase K (stored as a 20mg/ml stock solution at $-20°C$) to a final concentration of 100 μg/ml. Mix and incubate for 4 h at 37°C.

(vi)   Add NaCl to a final concentration of 0.1 M. Extract the mixture twice with an equal volume of buffer-saturated (redistilled) phenol. Recover the aqueous phase each time.

(vii)   Extract the aqueous phase once with an equal volume of 24:1 chloroform:isoamyl alcohol.

(viii) *Either* dialyse the aqueous phase extensively against TE buffer (1 mM EDTA, 10 mM Tris-HCl, pH 7.5)
        or add 2 volumes of ethanol, mix well and leave overnight at $-20°C$. Recover the DNA by centrifugation at 14 000 g for 15 min; air-dry the DNA pellet and redissolve it in TE buffer.

(ix)   Add RNase (pre-incubated at 90°C for 10 min) to a final concentration of 50 μg/ml. Incubate for 6 h at 37°C.

(x)   Repeat steps (vi)−(viii).

3.4 Preparation of Cloned DNA Probes

3.4.1 *Cloning Procedure*

The purity of any nucleic acid which is to be used as a hybridisation probe is of prime importance. If contaminating sequences are present, background hybridisation increases to a degree where interpretation of results is difficult, if not impossible.

If a probe sequence is derived from a small genome, such as a virus, a bacteriophage or a plasmid, purification is simple and yields are good since the desired sequence represents a significant proportion of the total nucleic acid, which itself is plentiful and easily separated from contaminating host DNA and RNA. For instance, 1 mg of a recombinant plasmid, 4 kb long, can be prepared from a few litres of bacterial culture and this will contain 250 μg of a potential probe gene 1 kb long.

As the size of the genome increases, the proportion of any 1 kb sequence decreases until it becomes negligible. The size of a mammalian genome, for instance, is typically 3×10^6 kbp. A culture of 5×10^8 tissue culture cells will yield up to 1 mg of DNA but will contain only 0.3 ng of any unique probe sequence 1 kb long. A similar problem is encountered with most eukaryotic mRNA species which represent only a low percentage of the total mRNA population. Here the problem is also compounded by the lability of mRNA.

The solution to these problems is to clone the probe sequences, that is, to insert a sequence or a mixture of partially-purified sequences containing the desired probe, into the DNA of a plasmid or bacteriophage vector. The usual cloning procedures are arranged such that each plasmid or phage DNA molecule takes up a single 'insert' sequence and, on transformation or infection of host bacteria, each takes up a single species of recombinant DNA. Thus colonies or plaques are produced each of which contains a unique plasmid or phage. The desired recombinant clone can be selected by its DNA, protein or genetic characteristics. Very little starting material is necessary for this process and the end result is a purified probe sequence propagated in a fast-growing bacterial system capable of high DNA yields.

The approach for mRNA sequences is the synthesis of double-stranded cDNA followed by cloning. The complete collection of cDNA clones thus obtained from total cellular mRNA or partially purified mRNA is known as a 'cDNA library'. The approach with genomic DNA is to cleave the DNA into fragments of suitable size by restriction endonuclease digestion followed by cloning to yield a 'genomic DNA library'. *Table 3* summarises the available vectors, their cloning sites, the sizes of DNA fragments which they will accept, and the primary recombinant selection methods. When choosing a cloning vector, the following points should be considered.

(i)   Bacteriophage lambda and plasmid pBR322 cross-hybridise to a certain degree. Therefore, do not plan to probe a library cloned using one of these vectors with a probe cloned into the other without first removing vector sequences from the probe.

(ii) It is easier to clone using plasmid vectors than using phage but the efficiency of transformation is $10^2 - 10^5$ times lower; many more transformants may have to be screened to identify the desired clone.

The main steps in a typical cloning experiment are as follows.

(i)   Isolate the DNA to be cloned.

(ii) Modify the ends by digestion with an appropriate restriction enzyme or by the addition of homopolymer tails or oligonucleotide linkers so that the fragment can be accepted by the vector of choice.

(iii) Cut the vector DNA with the appropriate restriction enzyme. Vector DNA cut with an enzyme which produces blunt ends will accept *any* blunt-ended insert. However, unless the blunt ends have been produced with the same restriction enzyme, a restriction site may not be regenerated after ligation of the vector and inserted sequences. In this case, it may not be possible to purify the inserted DNA away from the vector sequences after cloning.

(iv) Modify the vector to prevent self-ligation. This can be achieved for plasmid vectors by dephosphylating the 5′ ends of the vector with alkaline phosphatase *or,* for bacteriophages, by separating the bacteriophage arms from non-essential fragments where appropriate. This is unnecessary for insertion vectors or for those in which the primary selection of recombinants is based on the *spi* phenotype *(Table 3)*.

(v)   Ligate the DNA fragment to the vector DNA (this step is unnecessary when cloning cDNA, to which long homopolymer tails have been added, into complementary-tailed vectors, since the ends self-anneal sufficiently for ligation to take place in the bacterial host cell).

(vi) Introduce the recombinant DNA into a suitable host either by direct DNA transformation (suitable for both plasmids and phages) or by *in vitro* packaging the DNA into phage heads and infecting host bacteria (suitable for phage lambda vectors and cosmids only).

(vii) Plate the recombinants and perform primary screening to identify those clones carrying the sequences of interest.

Clearly, the detailed methodology of cloning is beyond the scope of this chapter. An excellent basic text for cloning procedures is that of Maniatis *et al.* (7) while more recent and advanced protocols have been described in other volumes of this series (1,2).

Table 3. Vectors for Cloning[a].

Vector type	Vector	Vector size (kb)	Cloning site	Insert size (kb) maximum (minimum)	Primary recombinant selection [b]	References
Plasmid	pBR322 [c]	4.3	*Eco*RI *Bam*HI *Pst*I	10 [d]	none (size) tet [s]/amp [r] amp [s]tet [r]	8
Plasmid	pAT153	3.7	as for pBR322	as for pBR322	as for pBR322	9
Plasmid	pBR325	5.9	*Eco*RI *Pst*I *Bam*HI *Ava*I	10 [d] 10 [d] 10 [d] 10 [d]	cam [s] tet [r] amp [r] amp [s] cam [r] tet [r] tet [s] cam [r] amp [r] none (size)	10
Plasmid	pBR328	4.9	as for pBR325	10 [d]	as for pBR325	11
Plasmid	pACYC184	4.0	*Bam*HI *Eco*RI	10 [d] 10 [d]	tet [s] cam [r] cam [s] tet [r]	12
Plasmid	pUC18/19	2.7	many [e]	10 [d]	inactivation of β-galactosidase gene (white colonies on X-gal medium)	13
Bacteriophage	M13 mp18/19	7.2	many [e]	5 [f]	as for pUC18/19 except blue plaques − white plaques	13
Bacteriophage	λgtwes.λB	40.4	*Eco*RI	15.1	non-recombinant DNAs do not package into λ heads (minimum genome size)	14
Bacteriophage	λ1059 [g,h]	44	*Bam*HI	24.4 (6.3)	spi [+] − spi [−] (growth in P2 lysogens)	15
Bacteriophage	L47.1 [h]	40.6	*Eco*RI *Hind*III *Bam*HI	21.4 (8.4) 19.8 (6.8) 17.2 (4.2)	spi [+] − spi [−] spi [+] − spi [−] spi [+] − spi [−]	16
Bacteriophage	Charon 28	39.39 [j]	*Bam*HI *Hind*III *Eco*RI	18.91 (5.96) 11.46 (0) 17.24 (4.24)	spi [+] − spi [− k] none packaging	17
Cosmid	pHC79	6.4	*Bam*HI *Pst*I *Eco*RI	40 (30) 40 (30) 40 (30)	tet [s] amp [r] amp[s] tet [r] none	18

[a] Reproduced with the permission of Bethesda Research Laboratories.
[b] amp [r/s] , ampicillin resistance/sensitivity; tet [r/s], tetracycline resistance/sensitivity; cam [r/s], chloramphenicol resistance/sensitivity.
[c] Specialised variants include: pKH47 and pHSV106.
[d] Plasmids can accept inserts larger than 10 kb but transformation efficiency and DNA yield decrease. High background is also a problem.
[e] *Hind*III, *Sph*I, *Pst*I, *Sal*I, *Acc*I, *Hinc*II, *Xba*I, *Bam*HI, *Xma*I, *Sma*I, *Kpn*I, *Sst*I, *Eco*RI, *Hae*III.
[f] Larger inserts have been cloned. Any fragment larger than 1 kb may be difficult to clone due to presence of *E. coli* K restriction system in hosts.
[g] Will not grow on recA [−] hosts.
[h] Can replicate as plasmid in λ lysogens.
[j] Stock 2221 (other stocks vary in length).
[k] Large variations in recombinant plaque size.

The use of hybridisation techniques for screening libraries and analysing recombinant DNA molecules is considered in depth in Chapter 5 of the volume.

3.4.2 *Removal of Repetitive Sequences*

The use of restriction enzymes (see ref.19 and Appendix I) enables the researcher to tailor DNA probes to his particular needs. A restriction map can readily be deduced for any cloned DNA fragment (3). The most appropriate portion of each fragment for use as a hybridisation probe can then be dissected from the whole. The choice of a piece of DNA for use as a probe depends upon the following factors:

(i)   Size; the larger the fragment, the greater the amount of radioactivity hybridised to homologous sequences, and hence the greater the sensitivity of their detection.

(ii)   The amount of unnecessary material it contains; the probe should be as specific as possible unless sandwich hybridisation techniques are being used or long single strand tails such as those generated with T4 polymerase (Section 4.1.3) or the M13 hybridisation probe primer (Section 4.2.1) are likely to be an advantage.

(iii)   The content of sequences which may cross-hybridise non-specifically with the DNA or RNA being probed.

The third factor is the most important. This must be considered very carefully, especially when genomic DNA forms part of any hybridisation mixture. In particular, the presence of repetitive sequences during Southern blot analysis (Chapter 5, Section 3.3) of genomic DNA can cause problems such as high background and masking of primary bands as well as hybridisation to secondary and unrelated bands.

A protocol for isolating a probe fragment of maximum size but with minimum repeat content is outlined below and shown diagrammatically in *Figure 1* (ref. 20).

(i)   Digest the cloned DNA which is to be used as the probe with a series of appropriate restriction enzymes.

(ii)   Separate the fragments by electrophoresis on a 1% agarose gel.

(iii)   Transfer the fragments to nitrocellulose by Southern blotting, taking duplicate nitrocellulose filters from the gel as described in Chapter 5, Section 3.3.

(iv)   Air-dry the filters on Whatman 3MM paper. Sandwich the filters between 3MM paper and bake at 80°C for 2 h.

(v)   Shear a sample of total genomic DNA by forcing it several times through a syringe needle (23-gauge).

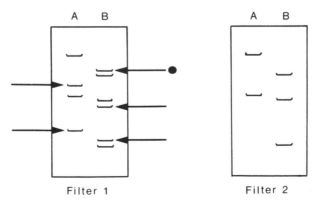

Figure 1. Diagrammatic representation of the method for detecting repetitive sequences. Two different restriction enzyme digests A and B are shown diagrammatically on each of filters 1 and 2. Fragments which contain no repeats (indicated by arrows) are those which appear on filter 1 but not on filter 2. The largest non-repetitive fragment, which is that most suitable for use as a probe, is indicated by ●. Details of the experimental procedure are given in the text.

(vi)   Separately nick-translate (Section 4.1.2) 0.5 μg of the sheared genomic DNA and 0.5 μg of intact cloned DNA.

(vii)   Hybridise one of the filters (Chapter 5) to the radioactive cloned DNA to visualise all the restriction enzyme fragments present in each digest (*Figure 1*, filter 1).

(viii)   Hybridise the other filter to the genomic DNA. Under these conditions, the only sequences which will be present at a concentration sufficiently high to hybridise to the immobilised DNA will be those which are highly repetitive (*Figure 1*, filter 2).

(ix)   Compare the autoradiographs of the two filters. This will allow identification of restriction fragments free of repetitive sequences, that is, those fragments in filter 1 that do not appear on filter 2 (*Figure 1*). The largest of these are the most useful as probes.

3.5 Isolation of Plasmid, Viral or Bacteriophage DNA

3.5.1 Plasmid DNA

Bacterial plasmid DNA can be isolated by the method given in *Table 4*. The aim of the protocol is to differentially precipitate high molecular weight (host cell) DNA and low molecular weight (supercoiled plasmid) DNA. The plasmid DNA is further purified *either* by caesium chloride equilibrium density gradient centrifugation in the presence of ethidium bromide *or* by chromatography through NACS resins (improved versions of the reverse phase resin RPC5; refs. 21,22). Both methods are described in *Table 4*. The NACS chromatography procedure is more labour intensive but much less time-consuming than the centrifugation method. The yield of plasmid DNA is comparable by either method.

3.5.2 Bacteriophage or Viral DNA

Linear double-stranded DNA. The method of preparation of linear, double-stranded DNA from bacteriophages such as phage lambda and viruses such as adenovirus type 2 consists of the following basic steps:

(i)   cell lysis;

(ii)   removal of cell debris;

(iii)   purification of virus or phage particles by caesium chloride density gradient centrifugation;

(iv)   DNA extraction from purified particles.

Detailed experimental protocols for the purification of bacteriophage and viral DNA are given in *Tables 5* and *6*, respectively.

RF bacteriophage and supercoiled viral DNAs. The double-stranded replicative form (RF) DNA of single-stranded bacteriophages such as M13 and ϕX174 can be isolated by the protocol given in *Table 4*. An essentially similar method is used for the isolation of the supercoiled DNAs of mammalian viruses such as SV40 or polyoma from infected cells (24).

3.6 Specific Probe Enrichment

Direct colony screening for recombinants containing DNA sequences (or cDNA copies) that are present in low abundance (<0.1% of total) is insensitive. The problem is often

Table 4. Preparation of Plasmid DNA[a].

1.	Grow the host bacteria in broth culture containing a suitable concentration of the appropriate antibiotic to which the plasmid confers resistance.
2.	Just before beginning the plasmid isolation, prepare lysozyme buffer: 50 mM glucose 10 mM EDTA 2−10 mg/ml chicken egg-white lysozyme 25 mM Tris-HCl, pH 8.0
3.	Harvest the bacteria by centrifugation at 4000 g for 10 min at 4°C.
4.	Resuspend the pellets in 4 ml of freshly prepared lysozyme buffer per 200 ml of original culture. Leave at room temperature for 10 min.
5.	Add 8 ml of 0.2 M NaOH, 1% SDS per 200 ml original culture, shake gently, and leave on ice for 10 min.
6.	Add 4 ml of 3 M potassium acetate, pH 4.8, per 200 ml original culture. Vortex thoroughly then leave on ice for 30 min.
7.	Centrifuge at 4000 g for 10 min at 4°C.
8.	Take the supernatant (which contains the supercoiled DNA), and add 0.6 volume of isopropanol. Mix well and leave at room temperature for 5 min.
9.	Harvest the precipitated DNA by centrifugation at 14 000 g for 10 min. Resuspend the pellet in 6.5 ml of TE buffer (10 mM Tris-HCl, pH 7.5, 1 mM EDTA) per 200 ml original culture.
10.	Purify covalently closed circular plasmid DNA *either* by equilibrium density gradient centrifugation in caesium chloride *or* by chromatography on NACS resin as follows.

Equilibrium density gradient centrifugation

1.	To each 6.5 ml solution of DNA in TE buffer, add 7 g CsCl and 0.5 ml of 5 mg/ml ethidium bromide solution (stored dark at 4°C).
2.	Centrifuge to equilibrium (36 h) at 120 000 g.
3.	At this time, the RNA will have pelleted onto the bottom of the tube and two DNA bands will be visible in the upper half of the gradient. Collect the more intense, lower plasmid DNA band through the side of the centrifuge tube using a syringe and needle. (The upper band contains nicked, open circles of plasmid DNA plus residual fragments of chromosomal DNA and is not collected).
4.	Remove ethidium bromide from the plasmid solution by repeated extractions with isoamyl alcohol saturated with CsCl solution.
5.	Dialyse the plasmid DNA preparation extensively against TE buffer.

Purification of supercoiled DNA by NACS chromatography

The resin of choice for this purpose is NACS 37 (BRL) which is suitable for use with a peristaltic pump. It has a capacity of 0.57 mg total nucleic acid per gram, and can be re-used repeatedly. RNA competes with supercoiled DNA for binding to this resin and so should be removed by RNase digestion prior to chromatography.

1.	To each 6.5 ml of plasmid solution in TE buffer (from step 9 of the first part of this table) add 200 units RNase T1 and incubate for 15 min at 37°C.
2.	Add NaCl to a final concentration of exactly 0.5 M. Mix well.
3.	Prepare a NACS column by suspending NACS resin in TE buffer containing 2 M NaCl, stirring gently overnight, pouring a column of suitable size (see capacity of NACS 37 resin above), and washing extensively with TE buffer containing 0.5 M NaCl.
4.	Load the plasmid DNA preparation from step 2 onto the column and wash with several column volumes of TE buffer containing 0.5 M NaCl.
5.	Elute the column with a linear gradient of 0.5−0.7 M NaCl in TE buffer and collect fractions. Monitor the A_{260} of the eluate. Residual RNA will be found in the loading eluate. Supercoiled plasmid DNA elutes in the gradient. Open circular plasmid DNA and genomic DNA fragments are retained on the column, which can be regenerated by extensive washing with 2 M NaCl in TE buffer followed by re-equilibration with loading buffer.
6.	Pool those fractions containing supercoiled DNA (as visualised by agarose gel electrophoresis) and add 2 volumes of ethanol. Place at −70°C for 30 min.
7.	Centrifuge at 12 000 g for 15 min. Resuspend the DNA pellet in TE buffer.

[a]From ref. 23.

Table 5. Purification of Bacteriophage Lambda DNA[a].

1.	Bacteriophage lambda-infected cells or induced lysogens are routinely lysed by adding one drop of chloroform for each $1-2$ ml of culture and incubating for 30 min at room temperature.
2.	To the lysate add 20 μl of 1 mg/ml stock solutions of RNase and DNase per litre of original culture. Centrifuge at 14 000 g for 10 min at 4°C to remove cell debris.
3.	Add 0.25 g CsCl per ml of phage solution. Layer onto a CsCl step gradient consisting of steps with densities (g/ml) of 1.7, 1.5 and 1.3 made up in 10 mM NaCl, 10 mM MgCl$_2$, 10 mM Tris-HCl, pH 7.6.
4.	Centrifuge at 125 000 g for 3 h at 4°C. Harvest the phage band.
5.	Adjust the CsCl density to 1.5 g/ml and re-centrifuge at 125 000 g for 22 h.
6.	Harvest the phage band and dialyse it extensively against 25 mM NaCl, 1 mM MgSO$_4$, 10 mM Tris-HCl, pH 8.0.
7.	Add EDTA to 10 mM and SDS to 0.2%. Heat at 65°C for 10 min.
8.	Add Proteinase K to 100 μg/ml and incubate at 37°C for 1 h.
9.	Add NaCl to 0.2 M and extract three times with buffer-saturated phenol[b] and once with chloroform:isoamyl alcohol (24:1 w/v).
10.	Dialyse the DNA extensively against TE buffer (10 mM Tris-HCl, pH 7.5, 1 mM EDTA).

[a]From ref.14.
[b]See Section 4.1.5, step (v).

Table 6. Purification of Viral DNA.

1.	Harvest the infected cells and wash with phosphate-buffered saline[a].
2.	Resuspend the cells in sterile 10 mM Tris-HCl, pH 7.9. Sonicate in a bath sonicator for two periods each of 1 min at 4°C to disrupt the cells.
3.	Add an equal volume of a fluorocarbon such as Freon 113 and blend in a homogeniser to solubilise membranes.
4.	Separate the phases by low speed centrifugation and collect the top (aqueous) phase.
5.	Layer this onto a CsCl step gradient composed of 7 ml of CsCl (density 1.4 g/ml) gently overlayed with 5 ml of CsCl (density 1.2 g/ml) both in 10 mM Tris-HCl, pH 7.4.
6.	Centrifuge at 125 000 g for 1 h. Virus particles collect at the interface and can be recovered through the side of the tube using a syringe and needle.
7.	Dilute the virus suspension at least 2-fold and layer onto pre-formed gradients made using equal volumes of the CsCl solutions prepared as described in step 5.
8.	Centrifuge at 125 000 g for 16 h and collect the virus band as in step 6. Dialyse extensively against TE buffer (10 mM Tris-HCl, pH 7.5, 1 mM EDTA).
9.	Add SDS to 1%. Mix well.
10.	Digest with Proteinase K and recover the viral DNA as described in *Table 5*, steps $8-10$.

[a]The composition of phosphate-buffered saline is 8.0 g NaCl, 0.2 g KCl, 1.5 g NaH$_2$PO$_4$, 0.2 g K$_2$HPO$_4$ per litre.

encountered when cDNA libraries from cells of two different tissues, different developmental stages or stages in tumour progression are being compared in an attempt to identify genes which are expressed in one cell type and not the other. To overcome this insensitivity, it is sometimes necessary to prepare a probe from which most of the mRNAs (or cDNAs) common to both cell types have been removed. This means that the mRNAs from genes expressed more in one cell type than the other will be specifically enriched. The preparation of the enriched probe is carried out by allowing sequences common to both cell types to anneal in a hybridisation reaction carried out either in solution or on a solid matrix to a chosen C_0t or R_0t value (see Chapter 1 for definition of these terms). The reannealed sequences are then separated from unannealed, single-strand sequences by hydroxyapatite chromatography (in the case of solution hybridisation) or by successive washing of the solid matrix. Protocols for both procedures are given below.

3.6.1. *Solution Hybridisation and Hydroxyapatite Chromatography*

(i) Calculate the C_0t or R_0t value necessary to remove the competing sequences.

(ii) Perform the reannealing reaction at appropriate DNA or RNA concentrations and for the time calculated to achieve the desired degree of hybridisation at 68°C in 0.12 M sodium phosphate buffer (or other conditions as determined by parameters described in reference 25).

(iii) After the required hybridisation time, dilute the reaction into 5 volumes of 0.4% SDS, 0.14 M sodium phosphate buffer, pH 6.8.

(iv) Make a slurry of hydroxyapatite in 0.4% SDS, 0.14 M sodium phosphate buffer, pH 6.8. Use this to prepare a suitable sized column of hydroxyapatite, water-jacketed at 60°C. (Each packed ml of hydroxyapatite is sufficient to bind 400 μg of DNA.)

(v) Add the diluted hybridisation mixture [step (iii)] to the column. Collect the eluate. Wash with 4 column volumes of 0.4% SDS, 0.14 M sodum phosphate buffer, pH 6.8, and pool this with the original eluate. This contains the single-stranded unannealed DNA.

(vi) Recover annealed DNA from the column by washing with 0.4% SDS, 0.4 M sodium phosphate buffer, pH 6.8.

(vii) Concentrate the single-stranded DNA by extraction with butan-2-ol as follows. Add an equal volume of butan-2-ol to the solution and mix the two phases throughly by vortexing. Spin the cloudy suspension briefly in a microcentrifuge, then discard the upper (butan-2-ol) layer.

(viii) Remove excess sodium phosphate by chromatography through Sephadex G50 in 1 mM EDTA, 50 mM NaCl, 10 mM Tris-HCl, pH 7.5

(ix) Repeat the entire procedure from step (i) as often as is necessary to achieve the desired enrichment (usually about 2−3 times).

3.6.2 *Hybridisation on a Solid Support* (26)

(i) Prepare the DNA or RNA which is to be bound to the solid support.

(ii) Couple this to *m*-amino-benzyloxymethyl-cellulose or to commercially-available epoxy- or triazine-activated celluloses (27,28) by methods described in reference 29 or on the manufacturers' data sheets.

(iii) Wash the cellulose five times with 20 volumes of water. (vortex, centrifuge and decant each time).

(iv) Block any unreacted groups on the cellulose by incubation with 20 volumes of 2 M ethanolamine, pH 9.0, at room temperature for 2−18 h.

(v) Wash the cellulose with water as in step (iii).

(vi) Immediately before use, wash the DNA cellulose six times with 90% formamide, 10 mM Hepes pH 7.0 (buffer A), then six times with *either* 0.6 M NaCl, 2 mM EDTA, 0.2% SDS, 10 mM Hepes pH 7.0 (buffer B) when using a DNA probe, *or* 0.6 M NaCl, 10 mM vanadyl ribonucleoside complex, 10 mM Hepes pH 7.0 (buffer C) when using an RNA probe.

(vii) Hybridise the DNA or RNA probe to the coupled cellulose for at least 18 h in buffer B or C, respectively, at 65°C using 1 ml microcentrifuge tubes mounted on a rotating shaker inside a 65°C oven. Keep the reaction volumes to a minimum (300 μl/100 mg cellulose).

(viii) Wash the cellulose nine times with buffer B for DNA probes or buffer C for RNA probes, to remove unhybridised nucleic acids which will be enriched for the desired probe sequences.

(ix)   Remove unwanted, hybridised sequences from the cellulose by washing nine times with buffer A. This regenerates the cellulose. The enriched sequences from step (viii) can be re-applied to the cellulose several times to achieve the desired degree of enrichment.

4. LABELLING NUCLEIC ACIDS FOR USE AS PROBES

4.1 Radioactive Labelling Methods

4.1.1 *Choice of Radioisotope*

The following radioactive isotopes are those most commonly used in nucleic acid hybridisations:

Isotope	Half-life	Type of decay	Energy
^{32}P	14.3 days	β	high
^{125}I	60.0 days	γ	medium
^{3}H	12.35 years	β	low

For nucleic acid hybridisation in solution and on filters, ^{32}P is the isotope of choice since its high energy results in short scintillation counting times and short autoradiographic exposures. Each of the four deoxyribonucleotides is available in an α-^{32}P-labelled form, of high or low specific radioactivity, suitable for incorporation into DNA using one of the polymerase reactions (Sections 4.1.2, 4.1.3). In addition, [γ-^{32}P]ATP is also available for 5' end-labelling RNA or DNA using polynucleotide kinase (Section 4.1.4). Similarly, [α-^{32}P]cordycepin triphosphate can be used for 3' end-labelling (Section 4.1.5).

The traditional isotope of choice for *in situ* hybridisation (Chapter 8) was ^{3}H since its low energy results in low backgrounds. However, a serious consequence of this low energy is very long time periods for autoradiographic exposure of slides. Therefore, ^{125}I is now often used instead. Nucleotides labelled with either isotope are commercially available for use in polymerase reactions.

4.1.2 *Nick Translation*

The process of 'nick translation' utilises DNase I to create single-strand nicks in double-stranded DNA. The 5' → 3' exonuclease and 5' → 3' polymerase actions of *E. coli* DNA polymerase I are then used to remove stretches of single-stranded DNA starting at the nicks and replace them with new strands made by the incorporation of labelled deoxyribonucleotides (30,31). As a result, each nick moves along the DNA strand being repaired in a 5' → 3' direction ('nick translation'). This technique is shown diagrammatically in *Figure 2*.

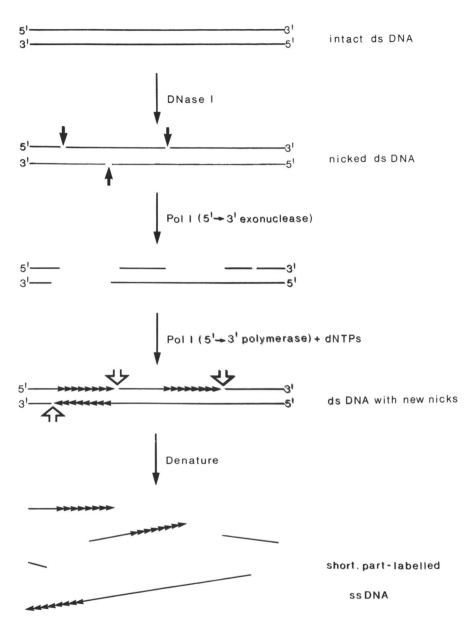

Figure 2. The preparation of probes by nick translation. ↑ , original nick position; ⇡ , final nick position; ▶▶▶, labelled strand; Pol I, *E. coli* DNA polymerase I; dNTPs, deoxyribonucleoside triphosphates; ds and ss DNA, double- and single-stranded DNA, respectively.

Nick translation can utilise any deoxyribonucleotide labelled with ^{32}P in the α position. [^{125}I]-, [^{3}H]- and non-radioactive biotinylated nucleotides can also be incorporated. With α-^{32}P-labelled nucleotides, final specific activities of 5 x 10^{8} d.p.m./μg DNA can be achieved. The detailed protocol is as follows.

(i)   To a microcentrifuge tube, add:
        0.5−1 μg DNA
        2 μl 0.05 mM dATP
        2 μl 0.05 mM dGTP
        2 μl 0.05 mM dTTP
        3.5 μl [α-^{32}P]dCTP (400−2000 Ci/mmol:10 mCi/ml)
        5 μl 10 x nick-mix (50 mM $MgCl_2$, 500 μg/ml BSA, 0.5 M Tris-HCl,
        pH 7.5)
        1 μl 3% 2-mercaptoethanol
        water to a final volume of 48 μl
        For higher specific activity probes, the unlabelled dTTP can be replaced with
        3.5 μl [α-^{32}P]dTTP (400−2000 Ci/mmol; 10 mCi/ml).

(ii)   Mix well; add 1 μl of DNase I (2 ng/ml) freshly diluted from a 2 mg/ml stock
        (stored at −20°C).

(iii)   Add 1 μl of DNA polymerase I (5 units). Incubate for 2 h at 16°C.

(iv)   Stop the reaction by adding 100 μl stop-mix (12.5 mM EDTA, 0.5% SDS,
        10 mM Tris-HCl, pH 7.5).

(v)   Chromatograph the mixture on a small Sephadex G-50 column (e.g., in a siliconis-
        ed Pasteur pipette) equilibrated in 3 x SSC. The DNA is excluded from the matrix
        and elutes ahead of the unincorporated deoxyribonucleotides. Collect 3-drop frac-
        tions from the column into microcentrifuge tubes and count directly by Cerenkov
        counting.

(vi)   Pool the fractions containing DNA.

The advantages of using nick translation as a labelling method are:

    (a) the simplicity of the reaction;
    (b) the uniform labelling of the DNA;
    (c) the high specific activities achieved.

The disadvantage is the nicking itself, which results in short single-stranded probe molecules in the hybridisation reaction.

An alternative protocol for nick translation is given in Chapter 8, *Table 3.*

4.1.3. *Labelling with T4 DNA Polymerase*

T4 DNA polymerase possesses two activities; a 3′→5′ exonuclease activity and a 5′→3′ polymerase (32,33). In the absence of exogenous deoxyribonucleoside triphosphates, only the exonuclease is active. It is more than 200 times as active as the exonuclease of DNA polymerase I and, under normal conditions, can remove 25 nucleotides per minute. When nucleotides are added, the polymerase activity predominates. The T4 DNA polymerase reaction is shown diagrammatically in *Figure 3*.

T4 polymerase will utilise any deoxyribonucleoside triphosphate labelled with ^{32}P in the α position or with ^{3}H or ^{125}I but has not yet been tested with biotinylated nucleotides. The detailed protocol is given below. Note that if the DNA to be labelled is a circular recombinant plasmid DNA, it must be first linearised by cutting with a suitable restriction enzyme.

(i)   At 0°C, dilute an aliquot of T4 DNA polymerase to 0.4 O'Farrell units/μl (33)
        using the storage buffer recommended by the manufacturer as the diluent. Mix
        well.

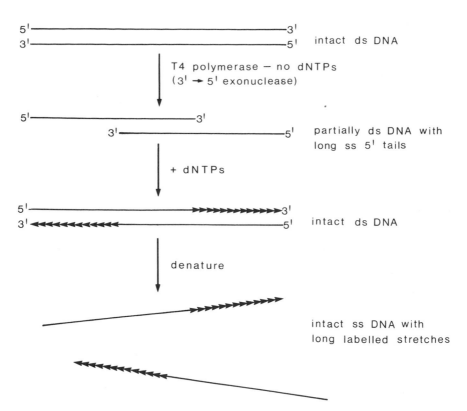

Figure 3. The preparation of probes using T4 DNA polymerase; ▶▶▶, labelled strand; dNTPs; deoxyribonucleoside triphosphates; ds and ss DNA, double- and single-stranded DNA, respectively.

(ii)   In another microcentrifuge tube, mix:
        2 μl 5 x reaction buffer (0.33 M sodium acetate, 50 mM magnesium acetate, 500 μg/ml BSA, 2.5 mM DTT, 0.165 M Tris-acetate, pH 7.9)
        0.2 μg target DNA
        4 μl sterile distilled water
        2 μl T4 DNA polymerase (0.8 units) as prepared in step (i).
(iii)   Mix, then centrifuge briefly to ensure the contents are at the bottom of the tube.
(iv)   Incubate at 37°C for the length of time needed to remove the required number of nucleotides (25 nucleotides removed per min under these conditions).
(v)   Cool the tube in ice water.
(vi)   Add to the tube:
        3 μl 5 x reaction buffer [see step (ii)]
        0.5 μl water
        2.5 μl 2 mM dCTP
        2.5 μl 2 mM dGTP
        2.5 μl 2 mM dTTP
        4.0 μl [α-^{32}P] dATP (400 Ci/mmol; 10 mCi/ml)
(vii)   Mix thoroughly and centrifuge briefly [as step (iii)]
(viii) Incubate at 37°C for 25 min.

(ix)   Stop the reaction by adding 5 μl of 0.3 M EDTA or heating at 65°C for 15 min.
(x)   Separate the labelled DNA from unincorporated nucleotides by chromatography
        through Sephadex G-50 as in Section 4.1.2, steps (v) and (vi).

The T4 DNA polymerase labelling procedure has several advantages over the nick
translation procedure.

(a) This procedure yields an intact double-stranded molecule with no nicks. It can
    therefore be cut with restriction enzymes and is more resistant to exonuclease
    attack.
(b) Defined regions of the DNA can be labelled by controlling the reaction condi-
    tions and thus the length of DNA labelled.
(c) DNA can be labelled to extremely high specific activity (10^9 d.p.m./μg) if high
    specific activity nucleotides (2000 Ci/mmol) are used. In this case, the volume
    of the [α-^{32}P]dATP used [step (vi)] must be increased to maintain the concentra-
    tion of dATP at the value given here so that the polymerase will work. The larger
    volume of [α-^{22}P]dATP required must be concentrated by drying *in vacuo* and
    then resuspended to 4 μl final volume ready for addition to the reaction mixture
    [step (vi)]. Although this yields highly radioactive DNA, DNA labelled to this
    level is unstable, that is, it is cleaved during the radioactive decay.
(d) Labelling of one strand of a double-stranded DNA fragment can be achieved.

The disadvantages of the method become evident when long stretches of DNA are to
be labelled, since secondary structures form in the long single-stranded regions. These
secondary structures inhibit the polymerase activity.

4.1.4 End-labelling with T4 Polynucleotide Kinase

Polynucleotide kinase is used to transfer the γ phosphate of ATP to a free 5′ OH group
in either DNA or RNA. The enzyme also has a phosphatase activity. Two reactions
are therefore possible (34,35). In the forward reaction, the enzyme catalyses
phosphorylation following removal of 5′-terminal phosphates with alkaline phosphatase.
In the exchange reaction, the kinase catalyses the exchange of an existing 5′ phosphate
with the γ phosphate of ATP. The latter reaction has to be carried out in the presence
of excess ATP and ADP if efficient phosphorylation is to be achieved. Both reactions
are most efficient with DNA which has a protruding 5′ terminus; recessed 5′ ends are
poorly phosphorylated. Specific activities of 5 x 10^5 c.p.m./pmol ends and 8 x 10^5
c.p.m./pmol ends can be achieved with the exchange and forward reactions, respectively, using blunt-ended DNA fragments. RNA is most effectively labelled by T4
polynucleotide kinase following base cleavage as described in *Table 7*.

The detailed protocols for end-labelling DNA or RNA with T4 polynucleotide kinase
are as follows.

Forward reaction.
(i)   Dissolve the DNA fragments (5 pmol of 5′ ends) in 100 μl of 50 mM Tris-HCl
        pH 8.0 and add 0.5 units of bacterial alkaline phosphatase. Incubate at 65°C
        for 60 min. Alternatively, for RNA fragments, use 1.5 units of alkaline
        phosphatase and incubate for 10 min at 45°C.

Table 7. Base Cleavage of RNA to be Labelled with T4 Polynucleotide Kinase.

1. To 200 μl of RNA at 50 μg/ml in distilled water in a microcentrifuge tube, add 20 μl of 1 M NaOH.
2. Leave on ice for 10 min to produce fragments of 200−1000 bases.
3. Add 20 μl of 1 M Tris-HCl pH 8.0 and sufficient HCl to neutralise the NaOH (check with pH paper).
4. Add 4 μl of 4 M NaCl plus 500 μl of ethanol. Leave at −20°C for at least 30 min.
5. Recover the precipitated RNA by centrifugation for 5 min in a microcentrifuge. Resuspend the pellet in distilled water at a concentration of 50 μg/ml. Store at −20°C.

(ii)   After the incubation, add stock EDTA to a final concentration of 10 mM. Heat at 65°C for 10 min then extract three times with an equal volume of buffer-equilibrated phenol.

(iii)   Add stock NaCl to a final concentration of 50 mM and precipitate the DNA with 2 volumes of ethanol at −20°C overnight or RNA with 3 volumes of ethanol at −20°C overnight.

(iv)   Re-dissolve the nucleic acid in 20 μl of 15 mM DTT, 10 mM MgCl$_2$, 0.33 μM ATP, 60 mM Tris-HCl, pH 7.8 containing 10 μCi [γ-^{32}P]ATP (>5000 Ci/mmol).

(v)   Add 5 units of polynucleotide kinase and incubate at 37°C for 30−60 min.

(vi)   Remove any unincorporated nucleotides by chromatography through G-50 Sephadex as in Section 4.1.2, steps (v) and (vi).

Exchange reaction.

(i)   Dissolve the DNA or RNA fragments (5 pmol of ends) in 25 μl of 12 mM MgCl$_2$, 1 mM DTT, 0.3 mM ADP, 0.5 μM ATP, 50 mM imidazole-HCl, pH 6.4.

(ii)   Add 50 μCi [γ-^{32}P]ATP (5000 Ci/mmol) and 5 units of T4 polynucleotide kinase.

(iii)   Incubate at 37°C for 30−60 min.

(iv)   Remove unincorporated nucleotides by chromatography through G-50 Sephadex [Section 4.1.2, steps (v) and (vi)].

The advantages of the use of T4 polynucleotide kinase to label nucleic acids for hybridisation are:

(a) it can be used to label RNA as well as DNA;

(b) the exact location of the labelled group is known;

(c) very small pieces of nucleic acid can be labelled;

(d) DNA fragments from a restriction enzyme digest can be labelled before separation by gel electrophoresis thus permitting the preparation of several probes at once.

The disadvantage is that the specific activity of the final product depends solely on the specific activity of the [γ-^{32}P] nucleotide used and cannot be adjusted by adding more enzyme.

4.1.5 End-labelling with Terminal Deoxynucleotidyl Transferase

Terminal deoxynucleotidyl transferase (often called terminal transferase) adds deoxyribonucleotides onto the 3′ ends of DNA fragments. Both single- and double-stranded DNAs are substrates for this enzyme, and if cobalt ions are present as co-factor, even recessed 3′ ends can be used as substrates.

Terminal transferase is useful in two ways:

(i)   to make complementary homopolymer tails on DNA fragments which are to be joined (by annealing and ligation) during a cloning experiment (36);

(ii)   to label the 3' ends of DNA molecules using [α-^{32}P]nucleoside triphosphates. Problems caused by the generation of homopolymer tails of random length can be avoided by the use of the nucleotide analogue [α-^{32}P]cordycepin triphosphate [3' deoxyadenosine 5' triphosphate (ref. 39)]. The lack of a 3' hydroxyl group on cordycepin triphosphate prevents further nucleotide additions, resulting in chain termination and the generation of DNA molecules labelled by the addition of a single nucleotide only.

The labelling procedure is as follows.

(i)   To the DNA (5 pmol of 3' ends), add 5 μl (50 μCi) [α-^{32}P]cordycepin triphosphate (>3000 Ci/mmol; 10 mCi/ml).

(ii)   Add 10 μl of 5 x tailing buffer (10 mM CoCl$_3$, 1 mM DTT, 0.5 M potassium cacodylate, pH 7.2).

(iii) Adjust the volume to 49 μl with water and mix well.

(iv)   Add 15 units of terminal transferase (1 μl). Mix and incubate at 37°C for 30 min.

(v)   Add an equal volume of phenol saturated with 50 mM Tris-HCl, pH 7.5, and mix to stop the reaction. Separate the phases by centrifugation for 5 min in a microcentrifuge.

(vi)   Transfer the upper (aqueous) phase to a new tube and add 0.5 volumes of 7.5 M ammonium acetate followed by 2 volumes of ethanol.

(vii) Store at −70°C for 15−30 min to precipitate the DNA.

(viii) Recover the DNA by centrifugation and redissolve the pellet in 50 μl TE buffer (10 mM Tris-HCl, pH 7.5, 1 mM EDTA).

(ix)   Repeat the ethanol precipitation [steps (vi)−(viii)] but air-dry the pellet and dissolve it in the buffer of choice. This procedure removes most of the unincorporated cordycepin. Alternatively the DNA can be purified by chromatography through Sephadex G50 [Section 4.1.2, steps (v) and (vi)].

The advantage of the use of terminal transferase to produce labelled DNA probes is that the molecules are specifically labelled at the 3' ends only. The disadvantage, as for T4 polynucleotide kinase, is that the specific activity of the final product is totally dependent on the specific activity of the [α-^{32}P]cordycepin triphosphate.

4.1.6 *End-labelling with the Large (Klenow) Fragment of E.coli DNA Polymerase I*

The large (Klenow) fragment of *E.coli* DNA polymerase I, which has the 5'→3' polymerase activity, can be used to 'fill in' the 3' ends of DNA fragments opposite naturally-occurring 5' extensions or those produced by some restriction enzymes e.g., *Bam*HI, *Hind*III (38). The reaction is shown diagrammatically in *Figure 4* and is described below.

(i)   To 1 μg of DNA in a common restriction enzyme buffer, such as 6 mM MgCl$_2$, 1 mM DTT, 50 mM NaCl, 6 mM Tris-HCl, pH 7.5, add 10 μl of a mixture of unlabelled nucleotides (0.1 mM each). The nucleotides supplied must be correct for the particular 3' end to be filled in, minus the one or more labelled

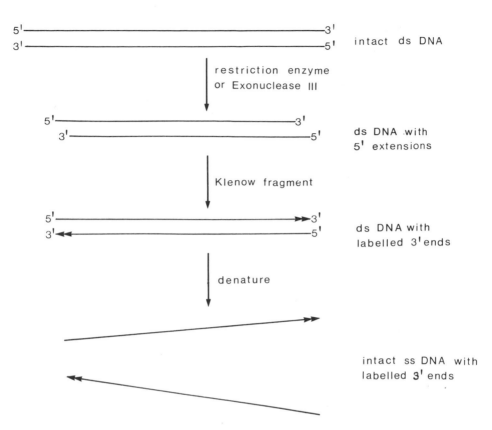

Figure 4. Preparation of probes using the large (Klenow) fragment of *E. coli* DNA polymerase I. ►► labelled strand; ds and ss DNA, double- and single-stranded DNA, respectively.

nucleotides to be used. For example, to fill in *Bam*HI 3′ termini, which are

CCTAG

G

add unlabelled dGTP, dTTP and dCTP if labelled dATP is being used. For very high specific activity probes, omit the unlabelled nucleotides and add only radiolabelled dGTP, dATP, dTTP and dCTP at step (ii).

(ii)   Add the α-^{32}P-labelled nucleotide(s) as aqueous solution(s) [2 μCi each (>3000 Ci/mmol)]

(iii)   Adjust the volume to 29 μl with water. Then add 1 μl (5 units) of the Klenow fragment of DNA polymerase I.

(iv)   Incubate for 15 min at room temperature.

(v)   Remove unincorporated nucleotides by chromatography through Sephadex G-50 [Section 4.1.2, steps (v) and (vi)].

The advantages of this method are:

(a)   it may be carried out directly after digesting DNA with a restriction enzyme; there is no need to purify the fragment(s)

(b) by adding the appropriate labelled nucleotides, all the 3' ends can be filled completely with stretches of labelled DNA, thus increasing the specific activity of the product.

The disadvantage is that it cannot be used on DNA with 3' overhanging termini. Blunt ends are labelled by the exchange of one base at the 3' OH terminus.

4.1.7 *The Use of Exonuclease III to Create 5' Overhangs in DNA Molecules*

If it is necessary to end-label DNA which has blunt ends or 3' overhangs, and replacement synthesis is desired, an alternative to T4 DNA polymerase labelling (Section 4.1.3) is to use exonuclease III to create recessed 3' ends which can then be filled in using either the Klenow fragment of DNA polymerase I or AMV reverse transcriptase as previously described (Sections 4.1.6 and 3.2, respectively). The reaction conditions for the use of exonuclease III are as follows:

> 5 mM $MgCl_2$
> 10 mM 2-mercaptoethanol
> $0.5-1$ μg DNA
> 50 mM Tris-HCl, pH 8.0
> Exonuclease III

The reaction with any particular DNA substrate should be optimised by titration using $1-4$ units of enzyme for $5-30$ min at 37°C.

4.2 **Special Cloning Systems for High Specific Activity Radiolabelling of Nucleic Acids**

4.2.1 *The M13 Universal Probe Primer*

High specific activity hybridisation probes can be generated using the M13 universal probe primer (39). The DNA which is to be used as probe should first be cloned into one of the 'even' series of bacteriophage M13 vectors (e.g., M13mp8) by standard techniques for cloning into plasmids (7). Do not use the 'odd' series of M13 vectors (e.g., M13mp9) since this may result in the production of a labelled probe which hybridises with pBR322-derived sequences. The 13-base sequence of the M13 universal probe primer ($^{5'}$GAAATTGTTATCC$^{3'}$) is complementary to the 5' side of the multiple cloning site of the family of M13 vectors and is used to initiate synthesis of the ($-$) strand from the ($+$) strand template by the Klenow fragment of *E.coli* DNA polymerase I. The synthesis of the complementary strand, which can be labelled by incorporation of an [α-^{32}P]deoxyribonucleotide, does not proceed to completion so that the inserted probe sequence remains single-stranded. The resulting hybridisation probe, at a specific activity of up to 5 x 10^8 d.p.m./μg, is therefore single-strand-specific and is used without denaturation for hybridisation analyses. The synthesis of the probe is shown diagrammatically in *Figure 5*.

The conditions used for making labelled probes by this method are much simpler than those used in more conventional labelling methods such as nick translation. A single reaction requires only 2 ng of primer and 50 ng of M13 ($+$) strand template containing the probe DNA insert, 10 μCi of [α-^{32}P]dNTP and one unit of the Klenow fragment of DNA polymerase I. A typical reaction is carried out as follows:

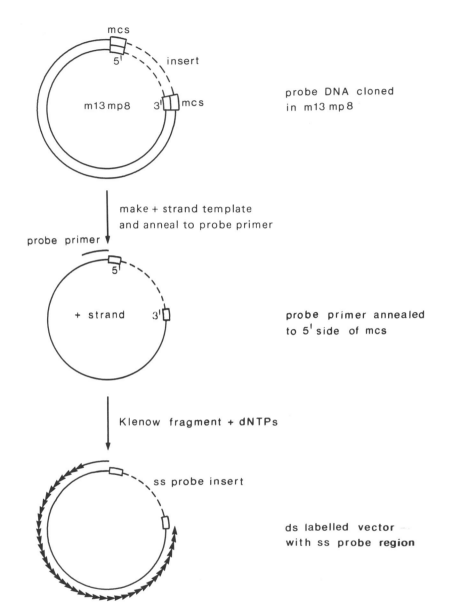

Figure 5. The use of the universal primer for labelling probes; MCS, multiple cloning site; ▶▶▶ , labelled DNA strand; dNTPs, deoxyribonucleoside triphosphates; ds and ss, double- and single-stranded DNA, respectively.

(i)   To a 0.5 ml microcentrifuge tube, add:
      2 μl universal primer at a concentration of 1 ng/μl
      1 μl M13 template DNA (50 μg/ml) containing the inserted probe sequence
      1.5 μl of a common 10 x restriction enzyme buffer
      6.5 μl water

(ii)   Mix well. Heat to 90°C to denature the DNA and then cool slowly to allow hybridisation of the primer to the template.

(iii)   To start the synthesis reaction, add the following components to the tube:

     1 μl 0.1 M DTT

     1 μl each 0.5 mM dCTP, dGTP, dTTP

     1 μl [α-^{32}P]dATP (400 Ci/mmol; 10 mCi/ml)

     1 μl Klenow fragment of DNA polymerase I (1 U/μl)

(iv)   Incubate at room temperature for 1 h.

(v)   Add 1 μl of 0.25 M EDTA pH 8.0 to stop the reaction. The probe is then ready for use. Note that the probe does not need to be denatured before use.

4.2.2 *The SP6 Polymerase System*

A dramatic increase in hybridisation efficiency can be achieved by using asymmetric probes in which only an mRNA coding strand is represented. These probes are best prepared using the SP6 cloning and transcription system (40,41). This consists of a pBR322 plasmid vector into which has been inserted the *Salmonella* phage SP6 RNA polymerase gene and its promoter. A few bases away from the promoter is a multiple cloning site into which a selected fragment of the DNA required as a hybridisation probe can be subcloned using standard methods (7). Transcription in the presence of [α-^{32}P]rNTP and SP6 polymerase proceeds from the SP6 promoter through the probe sequence, giving rise to an RNA transcript of high specific radioactivity and of the appropriate 'sense' for use as a hybridisation probe. The detailed protocol is as follows.

(i)   Once the recombinant plasmid has been prepared, linearise it with an appropriate restriction enzyme which cuts between the insert and the plasmid, leaving the insert attached to the SP6 promoter (see *Figure 6*).

(ii)   Add this DNA (0.5–1 μg) to an SP6 polymerase reaction mixture containing the following:

     6 mM MgCl$_2$

     10 mM DTT

     4 mM spermidine

     0.5 mM each ATP, GTP, CTP and UTP

     0.1–1.0 μCi [α-^{32}P]GTP (>400 Ci/mmol; 10 mCi/ml)

     40 mM Tris-HCl, pH 7.5

     60 units SP6 polymerase

     Final volume 50 μl

(iii)   Incubate the mixture at 37°C for 10 min.

(iv)   Stop the reaction by adjusting the reaction mix to 0.2 M sodium acetate pH 6.0; 10 mM EDTA, 0.1% SDS.

(v)   Purify the nucleic acid by phenol extraction [see Section 4.1.5, step (v)] and precipitation with 3 volumes of absolute ethanol at −20°C overnight.

(vi)   Remove the template DNA by digestion with RNase-free DNase I under conditions recommended by the manufacturer.

(vii)   Purify the SP6 polymerase transcripts by chromatography through Sephadex G150 in 1 mM EDTA, 50 mM NaCl, 10 mM Tris-HCl, pH 7.5, followed by phenol extraction and ethanol precipitation as in step (v).

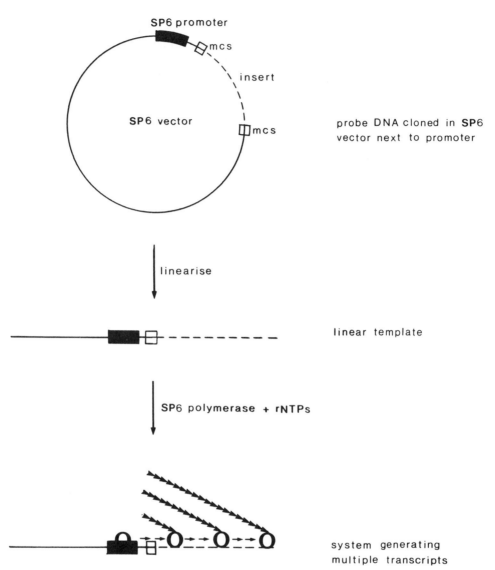

Figure 6. The SP6 polymerase system; MCS, multiple cloning site; ▶▶▶, labelled transcript; **O** , SP6 polymerase; ➔➔, direction of transcription; rNTPs, ribonucleoside triphosphates.

(viii) Dissolve the RNA pellet in sterile distilled water. It is then ready for use as a hybridisation probe.

The advantages of this system are:

(a) the versatility of the vectors;

(b) the ease of use of the SP6 polymerase (it is an extremely stable enzyme);

(c) the large amount of transcript produced (up to 10 μg from the 1 μg template);

(d) the high specific activity and intactness of the probes.

41

The only disadvantage seems to be the greater degree of care necessary for handling RNA as opposed to DNA. This system has proved invaluable for the production of probes for use during *in situ* hybridisation (42).

An alternative protocol for preparation of labelled RNA probes using SP6 polymerase is described in Chapter 8, *Table 2*.

4.3 Non-radioactive Labelling of Nucleic Acids

As previously mentioned, recent advances in nucleic acid technology have enabled researchers to develop non-radioactive labelling methods for the detection of DNA and RNA by hybridisation. Two methods predominate.

(i)   The incorporation of biotinylated nucleotides (e.g., bio-11-UTP) into DNA by standard techniques such as nick-translation. This biotin-labelled DNA is then used as the probe in a hybridisation reaction. The biotinylated DNA is detected by incubation with avidin or streptavidin which has been chemically coupled to an enzyme catalysing a colorimetric reaction (e.g., phosphatase, luciferase, peroxidase). The biotin binds the avidin-enzyme or streptavidin-enzyme complexes and the biotinylated DNA is then visualised by development of the colour reaction (43,44). As little as 2 – 5 pg target sequences can be detected by this method.

(ii)   The covalent linkage of an enzyme, such as those described in (i) above, to single-stranded DNA which can be used directly as a hybridisation probe. Again, the final stage of the procedure is the development of a colour reaction (45). Approximately 1 – 5 pg target sequences can be detected by this method.

Each of these methods is described in more detail below. The advantages of non-radioactive labelling and detection systems are obvious:

(a)   there is none of the hazard associated with the use of radioisotopes;
(b)   the probes are stable and their use is not restricted by a short half-life;
(c)   higher probe concentrations can be used to drive hybridisation reactions thus increasing sensitivity without causing high background problems;
(d)   detection is rapid (3 h) since no autoradiography is involved;
(e)   the reagents are inexpensive.

The main disadvantage seems to be the difficulty in recording the results by photographing the colour reaction products on nitrocellulose filters.

4.3.1 *Preparation and Use of Biotinylated DNA Probes*

(i)   Perform a nick-translation as described in Section 4.1.2. using 1 μg DNA and a final concentration of 20 – 25 μM bio-11-UTP or bio-16-UTP in place of dTTP. Allow the reaction to proceed for 90 min at 15°C. This results in the replacement of about one third of the T residues with bio-U.

(ii)   Purify the biotinylated DNA by chromatography on G-50 Sephadex [Section 4.1.2, steps (v) and (vi)] in the presence of 0.1% SDS which reduces the non-specific binding of biotin-labelled DNA to the microcentrifuge collection tubes.

(iii) Spot 1 μl of each fraction onto a sheet of nitrocellulose and detect biotin-containing fractions by the colour reaction [steps (vii) – (xviii)].

(iv)   The pooled fractions containing biotinylated probe can be stored at 4°C for several months. Such probes can be handled normally except that phenol extraction and the use of alkali for denaturation must be avoided.

(v)   Hybridise the biotinylated probes to immobilised DNA sequences by any standard method.

(vi)   Perform post-hybridisation washes by any usual method except that the final wash should be with 2 x SSC, 0.1% SDS.

(vii) It is not necessary to dry the filters for detection of sequences which have bound the probe. However, if they have been dried, rehydrate them in Triton buffer (0.1 M NaCl, 2 mM $MgCl_2$, 0.05% Triton X-100, 0.1 M Tris-HCl, pH 7.5). Rinse the wet filters in this solution.

(viii) Incubate the filters for 20 min at 42°C in 3% BSA.

(ix)   Gently blot each filter between sheets of filter paper and then dry the nitrocellulose in an 80°C oven for 10 min. The filters can be stored desiccated at this stage if desired.

(x)   Thoroughly rehydrate the filter in 3% BSA for 10 min and then drain it.

(xi)   Dilute streptavidin to 2 μg/ml (each 100 cm^2 filter needs 3 ml of this solution).

(xii) Incubate the filters in this solution for 10 min at room temperature. Decant the solution.

(xiii) Wash each filter in at least 30 ml of Triton buffer, three times.

(xiv) Dilute biotinylated polymer of alkaline phosphatase (available commercially or prepared as described in reference 44) to 1 μg/ml in Triton buffer (3 ml per 100 cm^2 filter). Incubate the filters in this for 10 min at room temperature. Decant the solution.

(xv)   Wash each filter twice with at least 30 ml of Triton buffer.

(xvi) Wash each filter twice in 0.1 M NaCl, 50 mM $MgCl_2$, 0.1 M Tris-HCl, pH 9.5.

(xvii) Incubate each filter in 7.5 ml alkaline phosphatase substrate solution (prepared as described in *Table 8*) in a sealed plastic bag, in the dark, for 2−4 h at room temperature.

(xviii) Wash the filters in 5 mM EDTA, 20 mM Tris-HCl, pH 7.5 to terminate the reaction. Examine the coloured DNA bands which will usually be most evident on one side of the filter. Filters may be stored dry or in sealed plastic bags containing this buffer. They must be stored in the dark.

4.3.2 *Preparation and Use of DNA Probes with Covalently-bound Enzyme* (45)

(i)   Dialyse 0.35 ml of calf intestinal alkaline phosphatase (9 mg/ml) against 0.1 M sodium phosphate buffer, pH 6.0

Table 8. Preparation of Alkaline Phosphatase Substrate Solution.

1.	Dissolve 5 mg of nitroblue tetrazolium (NBT) in 1.25 ml of TNM buffer (0.1 M Tris-HCl, pH 9.5, 0.1 M NaCl, 50 mM $MgCl_2$) and vortex for 1−2 min.
2.	Centrifuge briefly in a microcentrifuge and decant the supernatant into a plastic tube containing 10 ml of TNM buffer.
3.	Extract the NBT pellet three more times with 1.25 ml of TNM buffer and pool all the supernatants.
4.	Dissolve 2.5 mg of 5-bromo-4-chloro-3-indolylphosphate (BCIP) in 50 μl of dimethylformamide and add this solution drop-wise to the 15 ml of NBT solution. The solution is now ready for use.

(ii)   Add 90 μl of *p*-benzoquinone (30 mg/ml in ethanol) and incubate for 1 h at 37°C in the dark.

(iii)   Separate the phosphatase with covalently attached benzoquinone from unreacted benzoquinone by chromatography through G-25 Sephadex in 0.15 M NaCl. Pool the brown-coloured fractions (which contain the conjugated phosphatase).

(iv)   The next step is to cross-link the phosphatase to a synthetic polymer carrying many primary amino groups which will allow electrostatic binding to polyanionic DNA. To do this, raise the pH of the pooled fractions (~ 900 μl) by adding 100 μl of 1 M $NaHCO_3$ (to initiate cross-linking) and then add 20 μg Polymin G35, a polyethyleneimine.

(v)   Incubate in the dark at 37°C for 18 h.

(vi)   Dialyse against 5 mM sodium phosphate buffer, pH 6.8, and store at 3°C. This will be referred to as solution A.

(vii)   Linearise the probe DNA by digestion with an appropriate restriction enzyme. Denature 1 μg of the linearised probe DNA in 20 μl of 5 mM sodium phosphate pH 6.8 by heating at 100°C for 3 min. Transfer immediately to ice and leave there for 3 min.

(viii)   Add 20 μl of solution A [step (vi)] to the denatured probe.

(ix)   Add 6 μl of a 5% glutaraldehyde solution. Incubate for 10 min at 37°C. The probe is now ready to be used in any hybridisation reaction involving DNA immobilised on a nitrocellulose filter.

(x)   At the end of the hybridisation, wash the filters and develop the phosphatase colour as described in Section 4.3.1, steps (vii)−(xviii).

5. ACKNOWLEDGEMENTS

I have drawn much information from both unpublished data of colleagues at BRL, Gaithersburg, Maryland and BRL data sheets and publications. *Table 1* is reproduced with permission.

6. REFERENCES

1.   Glover,D.M., ed. (1985) *DNA Cloning − A Practical Approach*, Vol. **1**, IRL Press, Oxford and Washington D.C.

2.   Glover,D.M., ed. (1985) *DNA Cloning − A Practical Approach*, Vol. **2**, IRL Press, Oxford and Washington D.C.

3.   Smith,H.O. and Birnstiel,M.L. (1976) *Nucleic Acids Res.*, **3**, 2387.

4.   Burke,A.J. and Sharp,P.A. (1978) *Cell*, **14**, 695.

5.   Gait,M.J., ed. (1985) *Oligonucleotide Synthesis − A Practical Approach*, IRL Press, Oxford and Washington D.C.

6.   Smith,M. (1984) in *Methods of RNA and DNA Sequencing*, Weissmann,S.M. (ed.), Praeger Scientific, New York.

7.   Maniatis,T., Fritsch,E.F. and Sambrook,J. (1982) *Molecular Cloning. A Laboratory Manual*, Cold Spring Harbor Laboratory Press, New York.

8.   Bolivar,F., Rodriguez,R.L., Green,P.J., Betlach,M.C., Heyneker,H.L., Boyer,H.W., Crosa,J.H. and Falkow,S. (1977) *Gene*, **2**, 95.

9.   Twigg,A.J. and Sherratt,D. (1980) *Nature*, **283**, 216.

10.   Bolivar,F. (1978) *Gene*, **4**, 121.

11.   Soberon,X., Covarrubias,L. and Bolivar,F. (1980) *Gene*, **9**, 287.

12.   Chang,A.C.T. and Cohen,S.N. (1978) *J. Bacteriol.*, **134**, 1141.

13.   Norrander,J., Kempe,T. and Messing,J. (1983) *Gene*, **26**, 101.

14.   Leder,P., Tiemeier,D. and Enquist,L. (1972) *Science (Wash.)*, **196**, 175.

15. Karn,J., Brenner,S., Barnett,L. and Cesareni,G. (1980) *Proc. Natl. Acad. Sci. USA*, **77**, 5172.
16. Loenen,W.A.M. and Brammar,W.J. (1980) *Gene*, **10**, 249.
17. Rimm,D.L., Horness,D., Kucera,J. and Blattner,F.R. (1980) *Gene*, **12**, 301.
18. Hohn,B. and Collins,J. (1980) *Gene*, **11**, 291.
19. Roberts,R.J. (1984) *Nucleic Acids Res.*, **12**, r167.
20. Meuth,M. and Arrand,J.E. (1982) *Mol. Cell Biol.*, **2**, 1459.
21. Thompson,J.A., Blakesley,R.W., Doran,K., Hough,C.J. and Wells,K.D. (1983) in *Methods in Enzymology*, Vol. **100**, Wu,R., Grossman,L. and Moldave,K. (eds.), Academic Press, London and New York, p. 368.
22. Best,A.N., Allison,D.P. and Novelli,G.D. (1981) *Anal. Biochem.*, **114**, 235.
23. Birnboim,H.C. and Doly,J. (1979) *Nucleic Acids Res.*, **7**, 1513.
24. Hirt,B. (1967) *J. Mol. Biol.*, **26**, 365.
25. Wetmur,J. and Davison,N. (1968) *J. Mol. Biol.*, **31**, 349.
26. Scott,M.R.D., Westphal,K.-H. and Rigby,P.W.J. (1983) *Cell*, **35**, 557.
27. Moss,L.G., Moore,J.P. and Chan,L. (1981) *J. Biol. Chem.*, **256**, 12655.
28. Biagioni,S., Sisto,R., Ferraro,A., Caiafa,P. and Turano,C. (1978) *Anal. Biochem.*, **89**, 616.
29. Noyes,B.E. and Stark,G.R. (1975) *Cell*, **5**, 301.
30. Rigby,P.W.J., Dieckmann,M., Rhodes,C. and Berg,P. (1977) *J. Mol. Biol.*, **113**, 237.
31. Maniatis,T., Jeffrey,A. and Kleid,D.G. (1975) *Proc. Natl. Acad. Sci. USA*, **72**, 1184.
32. Morris,C.F., Hama-Inaba,H., Mace,D., Sinha,N.K. and Alberts,B. (1979) *J. Biol. Chem.*, **254**, 6787.
33. O'Farrell,P. (1980) *BRL Focus*, **3.3**, 1.
34. Richardson,C.C. (1965) *Proc. Natl. Acad. Sci. USA*, **54**, 158.
35. Maxam,A.M. and Gilbert,W. (1980) in *Methods in Enzymology*, Vol. **65**, Grossman,L. and Muldave,K. (eds.), Academic Press, London and New York, p.499.
36. Jackson,D.A., Symonds,R.H. and Berg,P. (1972) *Proc. Natl. Acad. Sci USA*, **69**, 2904.
37. Roy-Choudhury and Wu,R. (1980) in *Methods in Enzymology*, Vol. **65**, Grossman,L. and Muldave,K. (eds.), Academic Press, London and New York, p.43.
38. Downing,R.G., Duggleby,C.J., Villems,R. and Broda,P. (1979) *Mol. Gen. Genet.*, **168**, 97.
39. Hu,N. and Messing,J. (1982) *Gene*, **17**, 171.
40. Butler,E.T. and Chamberlain,M.J. (1982) *J. Biol. Chem.*, **257**, 5572.
41. Green,M.R., Maniatis,T. and Melton,D.A. (1983) *Cell*, **32**, 681.
42. Cox,K.H., DeLeon,D.V., Angerer,L.M. and Angerer,R.C. (1983) *Dev. Biol.*, **100**, 197.
43. Langer,P.R., Waldrop,A.A. and Ward,D.C. (1981) *Proc. Natl. Acad. Sci. USA*, **78**, 6633.
44. Leary,J.J., Brigati,D.J. and Ward,D.C. (1983) *Proc. Natl. Acad. Sci. USA.*, **80**, 4045.
45. Renz,M. and Kurz,C. (1984) *Nucleic Acids Res.*, **12**, 3435.

CHAPTER 3

Quantitative Analysis of Solution Hybridisation

BRYAN D. YOUNG AND MARGARET L.M. ANDERSON

1. INTRODUCTION

Nucleic acid hybridisation has been an essential technique for the development of our understanding of gene structure and function. The quantitative analysis of hybridisation has been used in many areas, but particularly in the measurement of genome complexity and gene copy number. The purpose of this chapter is to provide a framework for the interpretation of hybridisation in solution and to provide the theoretical basis of the computer program listed in Appendix III for the analysis of DNA-DNA and RNA-DNA annealing. The quantitative analysis of hybridisation to filter-bound nucleic acid is discussed in Chapter 4.

Nucleic acid reassociation involves the incubation of single-stranded molecules under conditions of ionic strength and temperature which allow the formation of base-paired duplex molecules. The simplest experiment which can be envisaged is the denaturation and subsequent reassociation (renaturation) of a pure, unbroken, perfectly base-paired DNA duplex. The initial event in the reassociation of a DNA duplex is a nucleation reaction between two complementary nucleotides of opposite strands. Many such events will take place until by chance the correct base pair is formed, which is followed by a rapid process of 'zippering' of the two single-stranded molecules into one duplex. A mathematical model of this process was originally developed by Wetmur and Davidson (1). It was concluded that the initial nucleation was the rate-limiting step and the zippering process was relatively fast. If the DNA strands are unbroken, each successful nucleation will result in a complete duplex molecule. However, unless the DNA to be reassociated is derived from a relatively small genome (e.g., viral DNA), it is necessary to shear the molecules to a size which allows reassociation to take place. When randomly-sheared DNA reassociates, duplex molecules with single-stranded tails will be formed. These tails will be free to participate in intermolecular reactions and eventually large concatameric molecules will be formed. The presence of single-stranded tails is an important factor when considering the measurement technique to be used. Some techniques, for example, hydroxyapatite chromatography (Section 7.2), measure the entire reassociated molecule, including single-stranded tails, as double-stranded. Other techniques, for example, resistance to digestion by the single strand-specific nuclease S1 (Section 7.3), measure only those parts of the molecule which are truly double-stranded.

If DNA is sheared, the concentration of fragments which contain a particular sequence is inversely proportional to the genome size. An important consequence of this is that the rate of the reassociation reaction between such strands is inversely proportional to the genome size. This was first observed by Wetmur and Davidson (1) and Britten and Kohne (2), who also found that the reassociation of DNA from eukaryotic cells was multi-component due to the presence of repeated sequences. In order to simplify the interpretation of DNA reassociation experiments, Wetmur and Davidson (1) introduced the notion of base sequence complexity, that is, the number of bases in non-repeated DNA sequence.

The relationship between base sequence complexity and reaction rate is fundamental to both DNA-DNA reassociation and to RNA-DNA hybridisation. An RNA preparation usually consists of relatively small molecules transcribed from a DNA template. Hence the base sequence complexity of an RNA population may be defined as the base sequence complexity of the DNA from which the RNA has been transcribed. The purpose of many nucleic acid reassociation experiments is to determine base sequence complexity by comparing the reaction rate of the nucleic acid of interest with that of a nucleic acid of known base sequence complexity. However, in order to interpret fully the experimental data, it is necessary to construct a mathematical description of nucleic acid reassociation. The simplest approach is to use the second-order reaction which describes a bimolecular reaction between two components, in this case the two complementary strands of a duplex. In the following analysis we explore the consequences of this model and demonstrate how it can be usefully applied to experimental data.

2. THE SECOND-ORDER REACTION

2.1 General Equation

Consider first the simple irreversible bimolecular reaction between two components whose concentrations are denoted by R and D and which react to form a product whose concentration is given by H, that is:

$$R + D \rightarrow H$$

We shall ultimately use R to represent the concentration of unhybridised RNA and D for the concentration of unhybridised DNA in hybridisation reactions. In DNA reassociation reactions we shall follow convention by using C to represent the concentration of DNA not in duplex form. The term H will represent duplex concentration, either RNA-DNA or DNA-DNA.

In a bimolecular reaction, the rate of product formation is given by:

$$\frac{dH}{dt} = k \times R \times D$$

At the beginning of the reaction ($t = 0$), the components have the concentration R_0 and D_0 and at a subsequent time t they have the concentrations $R = R_0 - H$ and $D = D_0 - H$. Hence:

$$\frac{dH}{dt} = k(R_0 - H)(D_0 - H) \qquad \text{Equation 1}$$

We may express H as a function of time by solving the following integral:

$$\int \frac{dH}{(R_0 - H)(D_0 - H)} = \int k dt + Q \qquad \text{Equation 2}$$

where Q is a constant.
Hence:

$$\frac{1}{D_0 - R_0} \cdot \log_e \frac{(D_0 - H)}{(R_0 - H)} = kt + Q \qquad \text{Equation 3}$$

At time $t = 0$, $H = 0$, and so:

$$Q = \frac{1}{(D_0 - R_0)} \cdot \log_e \frac{D_0}{R_0} \qquad \text{Equation 4}$$

Substituting for Q in Equation 3 and rearranging gives:

$$H = \frac{R_0 D_0 [1 - e^{(D_0 - R_0)kt}]}{R_0 - D_0 e^{(D_0 - R_0)kt}}$$

If the initial concentrations of the two reactants are not equal and we assume arbitrarily that $R_0 > D_0$, then it follows directly from Equation 4 that as $t \to \infty$ then $H \to D_0$. In other words, at the end of the reaction the concentration of the product is determined by the component which is at the lower initial concentration.

The general form of the second order reaction described in Equation 4 is applicable to nucleic acid reassociation where the two reacting components are assumed to be the two complementary strands which react to form a duplex molecule. However, two simplifications can be made to Equation 4. First, in an RNA-DNA hybridisation experiment one component is frequently in large excess over the other (i.e., $R_0 >> D_0$). Secondly in DNA-DNA reannealing the initial concentrations of the two components are often equal (i.e., $R_0 = D_0$). Both conditions lead to a considerable simplification of Equation 4.

2.2 $R_0 >> D_0$

If one component is in large excess over the other ($R_0 >> D_0$) then Equation 4 reduces to:

$$H = D_0 (1 - e^{-kR_0 t}) \qquad \text{Equation 5}$$

This is a pseudo first-order reaction and will adequately describe any second-order reaction in which one component is in an excess of greater than 20:1 over the other. The time course described by Equation 5 is illustrated in *Figure 1a* in the conventional manner as a semi-logarithmic plot. In hybridisation reactions it is convenient to express the reaction as a function of ($R_0 \times t$) where R_0 = initial RNA concentration. This eliminates any differences in reaction rate due to different RNA concentrations. In fact, such reactions are frequently characterised by the term $R_0 t_{1/2}$ which is equal to the value of $R_0 t$ for 50% of the reaction to have taken place. It can be shown from Equation 5 that $R_0 t_{1/2} = \log_e 2/k$ and hence measurement of $R_0 t_{1/2}$ is effectively a measurement of the rate constant k, provided that $R_0 >> D_0$.

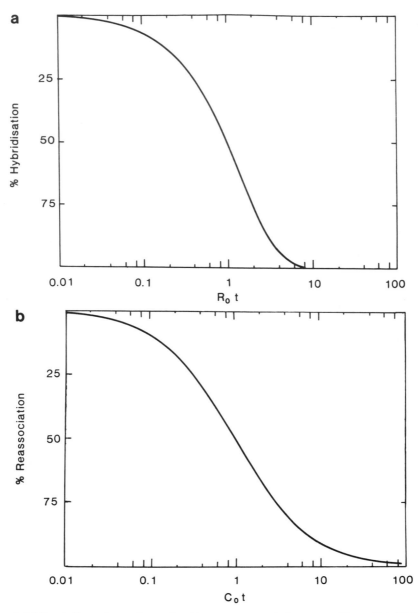

Figure 1. (a) Pseudo first-order reaction produced by one component in large excess ($R_o >> D_o$). **(b)** Second-order reaction with equal concentrations of starting components ($R_o = D_o$).

2.3 $R_o = D_o$

If the initial concentrations of both components are equal, the rate equation is:

$$\frac{dH}{dt} = k(D_o - H)^2 \qquad \text{Equation 6}$$

We may express H as a function of time by solving the integral

$$\int \frac{dH}{k(D_0 - H)} = \int kt + Q \qquad \text{Equation 7}$$

where Q is a constant.
Hence:

$$\frac{1}{D_0 - H} = kt + Q \qquad \text{Equation 8}$$

When $t = 0$, $H = 0$ and so $Q = 1/D_0$.
Hence Equation 8 can be rearranged to give:

$$H = D_0 \frac{kD_0 t}{kD_0 t + 1} \qquad \text{Equation 9}$$

In practice, this special case of the second-order reaction is usually applied only to the reassociation of DNA. In such a reaction the condition that $R_0 = D_0$ (equal concentrations of strand and anti-strand) is ensured if the DNA has previously been in its native form.

Following the convention adopted by Britten and Kohne (2), we will represent the total DNA concentration by C_0 and the concentration of single-stranded molecules by C. Equation 9 reduces to:

$$\frac{C}{C_0} = \frac{1}{1 + kC_0 t} \qquad \text{Equation 10}$$

DNA reassociation reactions are usually characterised by the term $C_0 t_{1/2}$ which is the value of $C_0 t$ for 50% of the reaction to have taken place. It can be shown from Equation 10 that $C_0 t_{1/2} = 1/k$. It can be seen from *Figure 1b* that this reaction is symmetrical when plotted on a semi-logarithmic scale, and occupies about 2 log units of $C_0 t$. This contrasts with the pseudo first-order reaction which occupies about 1.5 log units of $R_0 t$ and is asymmetric on a semi-logarithmic scale (*Figure 1a*).

3. APPLICATION OF SECOND-ORDER THEORY TO NUCLEIC ACID REASSOCIATION

The foregoing analysis applies to a simple bimolecular reaction between two components. This is limited in its application to most nucleic acid reassociation experiments since the experimenter often can measure only the total reaction time course which may represent a summation of a series of individual reactions. Despite such limitations this simple model has proven extremely useful in the interpretation of nucleic acid reassociation experiments as described below.

3.1 RNA-DNA Hybridisation

Our analysis of RNA-DNA hybridisation has assumed that DNA-DNA reassociation is either prevented by immobilisation on filters or does not take place because the DNA is single-stranded (e.g., single-stranded cDNA). However, if DNA reassociation can also occur then there will be two competing reactions. If similar concentrations of DNA and RNA are present, the reaction kinetics are rather complicated. However, if one or other component is in large excess the reactions are simpler. Galau *et al.* (3) studied both the hybridisation between excess (+) strand RNA with ϕX174 DNA and excess (+) strand DNA with ϕX174 DNA using hydroxyapatite chromatography. Care was taken to ensure that the reacting components were of similar size. Both the reactions followed pseudo first-order kinetics as described by Equation 5 and the rate constant of RNA-DNA hybridisation and DNA-DNA annealing differed by less than 25%. In contrast, when the RNA was reacted with an excess of double-stranded DNA, the rate constant of RNA-DNA hybridisation was 3- to 4.5-fold lower than that of DNA reassociation (4). This result holds true even if the RNA is reacted with excess complementary single-stranded DNA and therefore the difference cannot be explained by homologous DNA strands competing with the RNA in the hybridisation reaction. Therefore, the rate constant in DNA excess is 3- to 4.5-fold lower than in RNA excess. The reason for this anomalous effect is not known.

If double-stranded DNA is present in a large excess such that the formation of the RNA-DNA hybrids does not significantly reduce the DNA concentration, the kinetics of hybridisation can be analysed as follows (5). We define $k_H = k_D$ as the rate constants of RNA hybridisation and DNA reassociation, respectively. Equation 9 can be rewritten as:

$$\frac{H}{D_o} = 1 - \left(\frac{1}{k_D D_o t + 1} \right) \qquad \text{Equation 11}$$

Equation 1, which describes hybridisation kinetics, may be integrated separately if $D_o >> R_o$ and by substituting using Equation 11 we have:

$$\frac{H}{R_o} = 1 - \left[\frac{1}{(k_D D_o t + 1)^{k_H/k_D}} \right] \qquad \text{Equation 12}$$

It is apparent from Equation 12 that, if $k_H = k_D$, the rate of hybrid formation will equal that of DNA reassociation. If, as is more likely, $k_H < k_D$, the rate of hybrid formation will lag behind that of DNA reassociation. Melli *et al.* (5) showed that this effect can be quite dramatic. For example, if $k_H = 0.1 k_D$, the hybridisation reaction is 1000-fold slower than DNA reassociation as measured by $C_o t_{1/2}$.

3.2 DNA Reassociation

As described earlier, reassociation of randomly-sheared DNA involves formation of duplex molecules with single-stranded tails. This causes complications when estimating the extent of reassociation since hydroxyapatite (HAP) chromatography measures all parts of these molecules, including single-stranded tails, as double-stranded, whereas nuclease S1 and optical hypochromicity measure only the base-paired regions as double-stranded. Smith *et al.* (6) have exploited the difference between HAP chromatography and nuclease S1 degradation in an attempt to determine the effect of single-stranded tails on DNA reassociation. Experimentally it was found that reassociation of randomly-sheared *Escherichia coli* DNA, as determined by HAP chromatography, follows second-order kinetics, as described by Equation 10. However, when reassociation is followed by nuclease S1 degradation the reaction kinetics are not second-order, but can be empirically described by the equation:

$$\frac{C}{C_0} = \frac{1}{(1 + kC_0 t)^{-n}}$$

Equation 13

where the best fit value of n was found to be 0.45. Although this reaction is markedly slower than that measured by HAP chromatography, eventually all the DNA becomes resistant to nuclease S1. The nucleation rate of single-stranded tails was determined experimentally by reassociating a single-stranded tracer DNA with partially reassociated duplex molecules. It was found that the rate was reduced by a factor of 2 to 4.

In order to understand reassociation kinetics better, Britten and Davidson (7) constructed a computer simulation of the reassociation of randomly-sheared DNA fragments. Compared with such simulations, DNA reassociation, determined experimentally by both HAP chromatography and nuclease S1 degradation, exhibited a reduction in rate. The theoretical calculations of Britten and Davidson (7) suggest that this reduction can be accounted for by two factors. First, the reduction in rate is due to the decrease in average size of unreacted fragments. This is because larger fragments tend to react faster than smaller fragments. Secondly, a particle inhibition factor, due to the presence of a duplex region near a single stranded region, was postulated from theoretical calculations and found to be similar to that measured by Smith *et al.* (6). An important result of this analysis is that although Equations 10 and 13 empirically describe DNA reassociation, they to not offer much insight into the factors producing this rate reduction.

3.3 Reciprocal Plot Analysis of Hybridisation

In many of the early hybridisation experiments, reassociation of the DNA was prevented by binding the DNA to nitrocellulose filters. A large excess of labelled RNA was incubated with the filters and the quantity of RNA hybridised to DNA was assayed. This technique has the disadvantage that a relatively large volume is required and thus it is difficult to attain high values of $R_0 t$. Reciprocal plots were devised to provide reliable estimates of the rate constants and the final levels of hybridisation in experiments where completion of the reaction was difficult to obtain. The reciprocal plot can also indicate if the pseudo first-order reaction, Equation 5, can be re-written in the form:

$$\log \left[\frac{D_0}{D_0 - H} \right] = kR_0 t$$

Equation 14

Hence, if we plot $\log \left[\dfrac{D_0}{D_0 - H} \right]$ *versus* $R_0 t$ we obtain a straight line relationship the gradient of which is the rate constant k.

In practice, it is advantageous to fit the experimental data with a straight line using the least squares technique. Data which do not correspond to a second-order reaction will depart significantly from a straight line and it will be readily apparent that the model is inadequate.

3.3.1 *Double Reciprocal Plots*

It is often desirable to establish the final extent of a nucleic acid reassociation reaction.

In principle this can be accomplished by allowing the reaction to continue to high values of R_0t or D_0t (well past the $D_0t_{1/2}$, or $R_0t_{1/2}$) by either using high concentrations of the nucleic acid or long reassociation times. However, sometimes there are technical limitations to the upper values of R_0t or D_0t which are experimentally attainable. Double reciprocal plots have been used to circumvent this problem and so obtain reliable estimates of the final levels of hybridisation. Essentially, the reciprocal of the extent of hybridisation (H/D_0) is plotted *versus* the reciprocal of R_0t and D_0t. In this way an estimate of the final extent of hybridisation is obtained from the intersection of the plot with the ordinate (i.e., at R_0t or $D_0t = \infty$).

Consider the special case of $R_0 = D_0$. Equation 9 can be re-written in the form:

$$\frac{D_0}{H} = \frac{1}{kD_0t} + 1 \qquad \qquad \text{Equation 15}$$

Hence if D_0/H is plotted *versus* $1/D_0t$, the result is a straight line with a gradient of $1/k$ and an intercept of 1 on the ordinate. This predicts, as expected, that $H = D_0$ at $t = \infty$. Equation 15 indicates that such a double reciprocal plot can be used to plot data which conform to second-order kinetics with $R_0 = D_0$. However, Equation 5 describing a pseudo first-order reaction cannot be re-written in the double reciprocal form. In fact it can be shown, by returning to Equation 4, that a general second-order reaction would produce a curved double-reciprocal plot lying asymptotically between two intersecting straight lines (8). The equations of the two asymptotes are:

$$\frac{D_0}{H} = 1 \qquad \qquad \text{Equation 16}$$

$$\frac{D_0}{H} = \frac{1}{kR_0t} + 0.5 + 0.5 \frac{D_0}{R_0} \qquad \qquad \text{Equation 17}$$

This is illustrated in *Figure 2a* where it can be seen that for $R_0 = 2D_0$ the asymptote in Equation 17 cuts the ordinate at $D_0/H = 0.75$. If R_0 is very much greater than D_0 then the intercept is given by $= 0.5$. Hence a double reciprocal analysis which is based only on the early time points could overestimate the final level by as much as a factor of 2. This error could be reduced by obtaining values as close as possible to the ordinate. But of course, this simply means obtaining values at as high a R_0t or D_0t as possible, which in turn obviates the need for a double reciprocal plot.

A second form of double reciprocal plot which has been used is a plot of the reciprocal of hybrid concentration *versus* the reciprocal of RNA concentration. The purpose of this plot is to determine the quantity of hybrid which would be formed at an infinitely high RNA concentration. If a second-order reaction is displayed as a double reciprocal plot in the form D_0/H *versus* D_0/R_0, there are two asymptotes given by:

$$\frac{D_0}{H} = 1 \qquad \qquad \text{Equation 18}$$

and

$$\frac{D_0}{H} = \frac{D_0}{R_0}\left(\frac{e^z}{e^z - 1}\right) + \frac{(z-1)e^z + 1}{(e^z - 1)^2} \qquad \qquad \text{Equation 19}$$

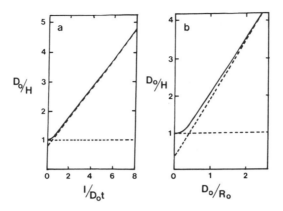

Figure 2. Double reciprocal plot of second-order reaction. (a) This plot was obtained from Equation 4 with $R_o = 2$, $D_o = 1$ and $k = 1$. The asymptotes (indicated by broken lines) were derived from Equations 16 and 17. (b) This plot was obtained from Equation 4 with $D_o = 1$, $k = 1$ and $t = 1$. The asymptotes were derived from Equations 18 and 19. Figure reproduced from ref. 8 with permission.

where $z = kD_o t$. This is illustrated in *Figure 2b* where it can be seen that, as in the previous double reciprocal plot, projection of the line through the data onto the ordinate will yield an overestimate of amounts of hybrid formed at infinite RNA and DNA concentrations.

Although such reciprocal plots have been of great assistance in the understanding of nucleic acid reassociation, they are becoming outmoded by more recent developments. Firstly, most reactions are now carried out in solution in small volumes, thus allowing high values of $R_o t$ and $D_o t$ to be obtained. Secondly, computer-based techniques of using least squares analysis to fit second-order reactions to experimental data have replaced such graphical methods.

4. MULTICOMPONENT REASSOCIATION

We have so far considered both DNA reassociation and RNA-DNA hybridisation to be represented by a single second-order reaction. This only applies to reactions in which all sequences are present in equal concentrations, for example, the reassociation of prokaryotic DNA. However, nucleic acids often consist of sequences which are not all present in equal concentration, for example, repeated and unique sequences in eukaryotic DNA. Hence the gross reaction can be distorted by the fact that those sequences present at higher concentrations will reassociate faster than those at lower concentrations. For this reason, both DNA reassociation and RNA-DNA hybridisation are often best treated as the sum of a number of independent second-order reactions, each of which has a different rate constant. The best method for such analysis is the non-linear least squares fitting procedure, which uses the algorithm described first by Marquadt (9) and in detail by Bevington (10). A microcomputer program based on this algorithm is listed and described in Appendix III. Programs to perform this task have also been published by Pearson *et al.* (11) and Kells and Strauss (12). In this approach the user defines in advance the number of components into which the data are to be analysed, and the program is used to find the rate constant and fraction of the total nucleic acid of each component yielding the best least squares fit. The user must supply initial estimates of each rate constant and fraction value and the program improves these estimates in an iterative manner until no further improvement is possible. The program of Kells and Strauss (12) has the feature that standard errors of the rate constants and fraction values are computed, and this feature has been incorporated into the program listed in Appendix III.

The number of components chosen by the user will depend on several factors. Firstly, the capacity and speed of the computer may be limiting. Secondly, a visual examination of the data may reveal a number of discrete components. If this is not the case, the user should try several analyses with an increasing number of components until no significant improvement in the degree of fit is obtained.

4.1 **DNA Reassociation**

The reassociation of prokaryotic DNA can be described by a single second-order reaction. This implies that there are no sequence repeats in such genomes. Experimentally it is found that the $C_0t_{1/2}$ values which characterise such reactions are proportional to the prokaryotic genome size. This is exactly the result expected from second-order analysis of DNA reassociation. Individual sequence concentration is inversely proportional to genome size, and so, therefore, is reaction rate as defined by the rate constant. Since $C_0t_{1/2} = 1/k$ it follows that $C_0t_{1/2}$ is proportional to genome size. Therefore, the genome size of an unknown DNA can be estimated by comparison of its $C_0t_{1/2}$ with that of a DNA standard whose genome size is known, e.g., *E. coli* DNA.

Unlike prokaryotic DNA, the reassociation of eukaryotic DNA is much slower and clearly consists of several kinetic components. The component which reassociates slowest represents sequences which are present in the genome only once and therefore is termed 'unique DNA'. As with prokaryotic DNA, the $C_0t_{1/2}$ of this component is proportional to genome size. Two components which reassociate faster are also usually observed. These are repetitive sequences and are typically referred to as repetitive and intermediate repetitive DNA. *Figure 3* shows the reassociation of the unique and intermediate repetitive components of human DNA. The curve through the points is the least squares fit of a two component analysis performed by the program described in Appendix III. In *Table 1* the $C_0t_{1/2}$ values and the fraction that each component represents of the total reacting sequences are presented along with the standard errors. The degree of repetition of the faster component can be estimated from this data by comparison of its $C_0t_{1/2}$ value with that of the unique DNA. Before this can be done, however, each $C_0t_{1/2}$ value has to be corrected for the fact that it refers to only a proportion of the DNA, that is, the observed $C_0t_{1/2}$ must be multiplied by the fraction of the total DNA represented by that particular component.

Another application of this type of analysis is determination of the repetition frequency of a particular gene. This can be done by following the reassociation of radioactive tracer DNA (complementary to the gene DNA) mixed with a large excess of total DNA. The repetition frequency is given by the ratio of the $C_0t_{1/2}$ value of the unique DNA to that of the tracer DNA.

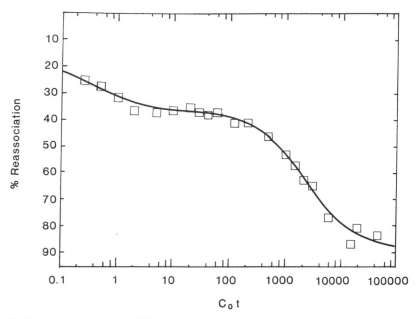

Figure 3. Reassociation of human DNA determined by hydroxyapatite chromatography. The fitted curve is the result of the analysis of the data into two components using the program listed in Appendix III. The results thus calculated are shown in *Table 1*.

Table 1. Computer Analysis of Reassociation of Human DNA[a].

Component[b]	Observed $C_ot_{1/2}$ (mol. nucleotide litre^{-1}. sec)	Fraction of total reacting sequences
1	0.45 ± 0.26[c]	0.176 ± 0.035
2	2148 ± 276	0.521 ± 0.016

[a]The data are from *Figure 3*.
[b]The data were resolved into two abundance components using the computer program described in Appendix III.
[c]The data are given $\pm$ S.E.

4.2 RNA-DNA Hybridisation

When native DNA is denatured and its reassociation studied, we can be certain that the concentration of each strand equals that of the anti-strand. In RNA-DNA hybridisation, however, the concentration of a particular DNA sequence need not be equal to the concentration of the complementary RNA sequence. Indeed, for given amounts of RNA and DNA there can be a complete range of relative concentrations at the sequence level, from RNA excess to DNA excess. If there is a wide range of different relative concentration for the sequences in the hybridisation reaction, the kinetics become extremely difficult to analyse. There is, however, one particular application of RNA-DNA hybridisation in which this complication does not arise; hybridisation of RNA to its cDNA copy. When reverse transcriptase is used to synthesise cDNA on a poly(A)$^+$ mRNA template, the relative concentration of particular cDNA molecules

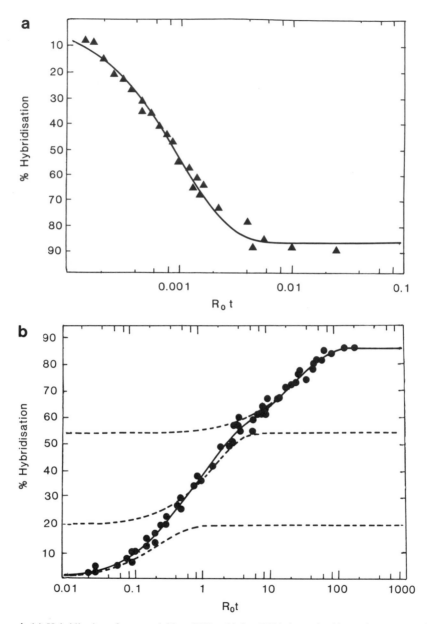

Figure 4. (a) Hybridisation of mouse globin mRNA with its cDNA determined by resistance to nuclease S1. The fitted curve is the result of the analysis of the data into one pseudo first-order component using the program listed in Appendix III. **(b)** Hybridisation of mouse myeloblast mRNA with its cDNA determined by nuclease S1 resistance. The fitted curves are the result of the analysis of the data into three pseudo first-order components and the results are presented in detail in *Table 2*. Data kindly provided by Dr. N.Affara.

reflects the concentrations of the equivalent mRNA molecules. Hence a 100:1 total mRNA to cDNA excess would result in a 100:1 RNA to cDNA excess at the sequence level. Such experiments are usually carried out under conditions of large RNA excess and therefore, from our earlier analysis, we can predict that the reaction will follow pseudo first-order kinetics as described by Equation 5.

If the mRNA preparation consists of a single messenger species, we can predict that the hybridisation of its cDNA in RNA excess will follow a single component pseudo first-order time course. In such experiments the cDNA is usually radioactively labelled and the reaction time course is followed by measuring the resistance of the cDNA-mRNA hybrid to degradation by nuclease S1. A typical example of this type of experiment is the hybridisation of purified mouse globin mRNA with its cDNA (*Figure 4a*). This reaction can be characterised by its $R_0t_{1/2}$ value which we have already shown equals $\log_e 2/k$. In an analogous manner to DNA reassociation, the $R_0t_{1/2}$ has been shown by Hell *et al.* (13) to be proportional to base sequence complexity of the RNA. If the RNA preparation consists of a single mRNA species, $R_0t_{1/2}$ will be proportional to the size of the mRNA molecule, assuming that it contains no repeated sequences. This holds true even if not all of the mRNA is copied into cDNA. If, on the other hand, the RNA preparation consists of a number of different mRNA molecules all present at equal concentration, then the complexity as measured by the $R_0t_{1/2}$ will simply be the sum of the lengths of different mRNA molecules. Hence, once a kinetic standard has been obtained, the hybridisation of a purified mRNA with its cDNA can be used to estimate the base sequence complexity and hence the purity of the mRNA.

In the mRNA-cDNA reactions discussed so far, the time course consists of a single pseudo first-order component. However, if different mRNA sequences are present at differing concentrations, then the mRNA-cDNA reaction time course will be multicomponent in nature. This is exactly analogous to the reassociation of eukaryotic DNA in which different degrees of repetition result in a multicomponent reaction. In this hybridisation reaction, however, different kinetic components are due to different concentrations or abundances of mRNA sequences. Such complex mixtures of mRNA molecules can be obtained from cells which are not committed to production of one protein to the exclusion of all others.

Unlike eukaryotic DNA reassociation, complex mRNA-cDNA reactions rarely show discrete kinetic components [although this has been observed by Bishop *et al.* (14)]. It is, therefore, usually necessary to resolve such reaction time courses using the non-linear least squares program described in Appendix III. An example of this type of experiment is shown in *Figure 4b*. It can be seen that although there were no visible discrete transitions between components, the program was able to compute the best least squares solution using three components. By comparison of the $R_0t_{1/2}$ values of these components with that of a known kinetic standard (e.g., globin mRNA-cDNA reaction), it is possible to calculate the base sequence complexity of each of the three components. Furthermore, it is possible to calculate the number of different types of mRNA sequence which are present in each component if the average weight of the mRNA has been measured or is assumed (*Table 2*). The slowest component is the most complex and the fastest is the least complex. The total complexity is given by the sum of the values of each kinetic component. In the absence of discrete transitions in the data, the experimenter should try analysing the data into an increasing number of components until no significant improvement in the degree of fit is obtained. By analysing such data into a small number of components, the experimenter may well be approximating to a time course which in reality consists of a large number of small overlapping

Table 2. Computer Analysis of RNA-DNA Hybridisation[a].

Component[b]	Observed $R_o t_{1/2}$ (mol. nucleotide litre^{-1}. sec)	Fraction of cDNA	Fraction of hybridisable cDNA	Corrected $R_o t_{1/2}$[c]	No. of different mRNA species[d]
1	0.16	0.199	0.232	0.03712	37
2	1.01	0.343	0.400	0.404	407
3	19.9	0.315	0.367	7.3	7372

[a]The data are from *Figure 4b*.
[b]The data were analysed into three components using the least squares fitting program described in Appendix III.
[c]Observed $R_o t_{1/2}$ values (first column) were corrected for the proportion of cDNA represented by each component.
[d]The globin mRNA-cDNA reaction was used as a kinetic standard in the analysis ($R_o t_{1/2}$, 6.6 x 10^{-4}; mol. wt. 4 x 10^5). The mouse myeloblast mRNA was assumed to have a molecular weight of 6 x 10^5.

components. Thus the abundance classes calculated by the computer may represent approximations of a continuum of mRNA abundance.

5. ANALYSIS OF RNA-cDNA TITRATION CURVES

Complementary DNA to a particular mRNA can be used as a probe to determine the concentration of the mRNA in any RNA preparation. There are two ways in which this can be done. First, the time course of the reaction of the cDNA with an unknown RNA can be compared with that of the cDNA with its own mRNA, and the fraction of the RNA which is the mRNA is given by the ratio of the $R_o t$ values of the two reactions. However, this approach relies on reacting the cDNA with excess mRNA and since the proportion of mRNA in the preparation is unknown it may be that an insufficient RNA/cDNA ratio is used. An alternative approach, which does not require a large mRNA/cDNA excess, is to titrate the cDNA with increasing amounts of RNA. In this method a fixed amount of cDNA is reacted with various amounts of RNA and the fraction of cDNA hybridised is plotted as a function of the RNA/cDNA ratio. Two idealised titration curves are plotted in *Figure 5*. Curve A is produced by assuming that the cDNA is complementary to 100% of the RNA, and curve B assumes that the cDNA is complementary to 10% of the RNA. The ratio of the slopes gives an estimate of the fraction of the RNA which is complementary to the cDNA. In practice the titration with an unknown RNA is always compared with that of the original mRNA. This is necessary because, unlike the theoretical curves shown in *Figure 5*, experimentally the approach to 100% hybridisation never gives a sharp inflection point. It is therefore difficult to determine the ratio at which complete hybridisation has been obtained.

An essential requirement of this approach is that sufficient time is allowed at all RNA/cDNA ratios for completion of the hybridisation reaction. Complementary DNA is single-stranded and therefore the kinetics of its reaction with RNA may be analysed in terms of a simple second-order reaction (15). We may therefore apply Equation 4 when the RNA and cDNA concentrations are not equal, and Equation 9 when they are equal. We shall characterise the reaction in terms of $D_o t_{1/2}$ (where D_o = initial cDNA concentration and $t_{1/2}$ = time for 50% of the reaction to take place). This yields

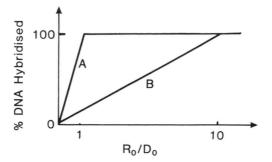

Figure 5. Idealised titration curves. **(A)** Titration of cDNA with complementary RNA. The cDNA is complementary to 100% of the RNA. **(B)** Titration of cDNA with RNA of which only 10% is complementary.

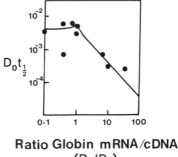

Ratio Globin mRNA/cDNA
(R_0/D_0)

Figure 6. Experimental variation of $D_0t_{1/2}$ with globin mRNA:cDNA ratio. The superimposed curve is theoretically derived, assuming that the cDNA-mRNA reaction has a rate constant of 182 litre. mol.$^{-1}$ sec^{-1} (from ref. 13 with permission).

the following relationships:

When $R_0 > D_0$,
$$D_0t_{1/2} = \frac{\log\left(2 - \frac{D_0}{R_0}\right)}{\left(\frac{R_0}{D_0} - 1\right)}$$ Equation 20

$R_0 = D_0$,
$$D_0t_{1/2} = \frac{1}{k}$$ Equation 21

$R_0 < D_0$,
$$D_0t_{1/2} = \frac{\log\left(2 - \frac{R_0}{D_0}\right)}{\left(1 - \frac{R_0}{D_0}\right)k}$$ Equation 22

The values of $D_0t_{1/2}$ can be determined experimentally for different values of R_0/D_0. These have been plotted in *Figure 6* for the reaction of mouse globin mRNA with its cDNA. The theoretical curve which is predicted by Equations 20 – 22 is superimposed on the data, assuming that $k = 182$ litre. mol^{-1}. sec^{-1}. It is clear from *Figure 6* that the slowest reaction, which has a $D_0 t_{1/2}$ of 5.5 x 10^{-3} mol. litre^{-1}. sec, occurs for $R_0 = D_0$. Thus if all reactions are taken to a $D_0 t$ value of at least 20 times this value, we can be sure that all reactions are complete irrespective of the ratio R_0/D_0. Hence, provided the rate constant of the reaction between the cDNA and its complementary mRNA is known, conditions can be chosen such that all reactions between a cDNA and an RNA whose complementary sequence content is unknown are complete.

6. FACTORS AFFECTING REACTION RATES

The rate at which complementary strands of nucleic acid form stable base-paired duplexes is dependent on a number of factors. These can broadly be divided into factors concerning the state of the nucleic acids, the conditions of incubation and the methods of detection used. If a set of reannealing or hybridisation experiments is carried out under identical conditions using nucleic acids of identical composition then the rates of reactions may be compared directly. If, however, as is more often the case, there are differences in the condition or composition of the nucleic acids, it is important to know which correction factors should be applied before reaction rates may be compared. The influence of such factors has been extensively reviewed elsewhere (16 – 18) and so only the main conclusions are presented here. Additional discussion of some of these factors is given in Chapter 4.

6.1 Nucleic Acid Length

The theoretical considerations of Wetmur and Davidson (1) predicted that the rate of reannealing of DNA would be proportional to the length (L) of the fragments. However it was found experimentally to be proportional to the square root of the length. This difference was attributed to an excluded volume effect which results in the complementary strands being unable to interact completely. The dependence on L was shown to hold over three orders of magnitude starting with L = 100 nucleotides.

In most DNA reannealing experiments the average length of the complementary strands is the same since DNA is usually randomly sheared prior to annealing. In some experiments, however, DNA strands of different lengths may be reannealed and it has been shown that the rate of such a reaction is proportional to the square root of the shorter of the annealing strands (19).

Hutton and Wetmur (20) studied rates of hybridisation using ϕX174 DNA and RNA transcribed with *E. coli* polymerase. Their results established clearly that the rate of hybridisation was proportional to the square root of length of the RNA, which in this case was the shorter strand. These results hold over a 5-fold range in molecular weight. Hence it has been well established that the rate of both DNA reannealing and RNA-DNA hybridisation is proportional to the square root of the length of the shorter strand.

6.2 Base Composition

Hutton and Wetmur (20) demonstrated that if the genome size and molecular weight of sheared DNA is taken into account, as indicated by Wetmur and Davidson (1), bacteriophage T2 DNA (34% G + C) and bacteriophage T7 DNA (50% G + C) have

Table 3. Variation of DNA Reassociation Rate with Ionic Strength[a].

Sodium phosphate buffer, pH 7.0 (M)	Monovalent cation conc. (M)	Relative reassociation rate[b]	Sodium phosphate buffer, pH 7.0 (M)	Monovalent cation conc. (M)	Relative reassociation rate[b]
0.01	0.015	0.0000	0.31	0.465	3.8841
0.02	0.030	0.0016	0.32	0.480	4.0085
0.03	0.045	0.0133	0.33	0.495	4.1300
0.04	0.060	0.0453	0.34	0.510	4.2487
0.05	0.075	0.1021	0.35	0.525	4.3646
0.06	0.090	0.1831	0.36	0.540	4.4778
0.07	0.105	0.2858	0.37	0.555	4.5884
0.08	0.120	0.4063	0.38	0.570	4.6964
0.09	0.135	0.5410	0.39	0.585	4.8019
0.10	0.150	0.6867	0.40	0.600	4.9049
0.11	0.165	0.8404	0.41	0.615	5.0056
0.12	0.180	1.0000	0.42	0.630	5.1040
0.13	0.195	1.1633	0.43	0.645	5.2001
0.14	0.210	1.3288	0.44	0.660	5.2941
0.15	0.225	1.4954	0.45	0.675	5.3860
0.16	0.240	1.6619	0.46	0.690	5.4758
0.17	0.255	1.8277	0.47	0.705	5.5636
0.18	0.270	1.9920	0.48	0.720	5.6495
0.19	0.285	2.1544	0.49	0.735	5.7335
0.20	0.300	2.3146	0.50	0.750	5.8157
0.21	0.315	2.4722	0.55	0.825	6.2009
0.22	0.330	2.6271	0.60	0.900	6.5478
0.23	0.345	2.7791	0.65	0.975	6.8613
0.24	0.360	2.9280	0.70	1.050	7.1457
0.25	0.375	3.0739	0.75	1.125	7.4047
0.26	0.390	3.2167	0.80	1.200	7.6413
0.27	0.405	3.3563	0.85	1.275	7.8582
0.28	0.420	3.4929	0.90	1.350	8.0575
0.29	0.435	3.6263	0.95	1.425	8.2412
0.30	0.450	3.7567	1.00	1.500	8.4110

[a]Adapted from ref. 17.
[b]This is the reassociation rate relative to that in 0.12 M sodium phosphate buffer, pH 7.0.

the same rate constant of reannealing. It therefore appears that, provided the genome size and average molecular weight of DNA is known, the effect of base composition on rate of annealing can be discounted.

There has been no detailed investigation of the effect of base composition on the rate of RNA-DNA hybridisation. Birnsteil *et al.* (21) measured the rates of hybridisation of different rRNAs to DNA immobilised on nitrocellulose filters and concluded that the slight differences between actual and predicted rates might be due to differences in base composition. However, Bishop (22) compared hybridisation reactions in solution between *E. coli* DNA (50% G + C) and its cRNA and between *P. mirabilis* DNA (38% G + C) and its cRNA, and showed that in 2 × SSC at 70°C the reaction rates were almost identical (SSC = 0.15 M NaCl, 0.015 M trisodium citrate). Hence we may conclude that the base composition of nucleic acids has little effect on the rate of DNA-DNA annealing or RNA-DNA hybridisation.

6.3 Ionic Strength

The effect of ionic strength on DNA reannealing was first studied by Wetmur and Davidson (1) who showed that the rate was strongly dependent on sodium ion concentration at least up to 3.2 M. At concentrations up to 0.2 M the rate has been shown to be proportional to the cube of the ionic strength (23). Comparisons of reaction rates at different salt concentrations can be greatly simplified by use of relative reaction rates which Britten *et al.* (17) computed from an empirical formula (see *Table 3*). It should be noted that the original data on which this table was based were obtained at 25°C below the melting temperature of DNA at each salt concentration. Hence the relative rates of reaction in this table represent optimal rates and may not apply directly to reactions which are carried out at sub-optimal temperatures. Furthermore the formula applies to measurements of DNA reannealing using hydroxyapatite for analysis. However, the table will still apply to measurements of relative rates.

The effect of ionic strength on rate of RNA-DNA hybridisation has not been studied in detail. However, it does appear that RNA-DNA hybridisation is also strongly dependent on ionic strength. For example, Nygaard and Hall (24) demonstrated a 5- to 6-fold increase in the rate of hybridisation of phage T2 RNA to T2 DNA by increasing the ionic strength from 0.2 M to 1.5 M NaCl.

6.4 Viscosity

In considering the effect of viscosity on the rate of reaction it is important to distinguish between microscopic and macroscopic viscosity. Microscopic viscosity refers to the micro-environment around the DNA bases and is commonly altered by the addition of sucrose, glycerol or sodium perchlorate. The macroscopic viscosity is dependent on the presence of polymers (including DNA) which will have no effect on the micro-environment. Thus a measurement of viscosity with a viscometer will include both micro- and macroscopic viscosity if there are polymers present.

Thrower and Peacocke (25) and Subirana and Doty (26) observed that increasing the microscopic viscosity with sucrose decreased DNA renaturation rates. A more detailed analysis (1) showed that in sucrose, glycol, ethylene glycol and sodium perchlorate the optimal rate of DNA renaturation is inversely proportional to the microscopic viscosity. This observation of the effect of microscopic viscosity has been confirmed by Chang *et al.* (27) who extended their studies to include the effect of macroscopic viscosity on DNA renaturation. It was shown that in a 5.7% Ficoll solution (a neutral polymer) the rate of phage T4 DNA renaturation was increased by 50%. Similarly in a 2% solution of the anionic polymer, sodium dextran sulphate, the rate of phage T2 DNA renaturation was increased by 4-fold, despite a 6-fold increase in the macroscopic viscosity. This increase in reaction rate was attributed to the exclusion of DNA from a substantial volume of the solution by the added polymers, resulting in an increase in the effective DNA concentration. Dextran sulphate is now widely used to accelerate reactions to filter-bound DNA (see Chapter 4).

6.5 Denaturing Agents

The optimal temperatures for nucleic acid reassociation in aqueous salt solutions lie in the range 60 – 75°C. However, the extended incubation at such temperatures which is required for reassociation of complex eukaryotic nucleic acid can lead to a considerable amount of thermal strand scission. Hence it is desirable to reduce the temperature whilst maintaining the stringency of the nucleic acid interaction. This can be achieved by introducing a reagent, such as formamide, which destabilises double-stranded nucleic acid. Thus, a 1% increase in formamide concentration lowers the T_m of native duplex DNA by 0.72°C. In contrast, for RNA-DNA hybrids the relationship between formamide concentration and the depression of T_m is not linear (see Chapter 4, section 4). However, altered conditions generally result in a reduction in reaction rate and therefore the experimenter must decide whether the benefit of a reduced rate of degradation outweighs the disadvantage of a reduced reaction rate. Although formamide is the most popular and well characterised reagent used to reduce reaction temperatures (28,29), a variety of other denaturing solvents have been tried including ethylene glycol (1), sodium perchlorate (27), tetramethylammonium chloride and tetraethylammonium chloride (27) and urea (17,30).

The most systematic study of the effect of formamide on the reassociation rate of DNA was carried out by Hutton (30). It was shown that the characteristic bell-shaped dependence of renaturation rate on temperature is maintained in solutions containing formamide. Furthermore, increasing the formamide concentration decreases the optimal renaturation rate by 1.1% per 1% of formamide. It was also demonstrated that the reduction in rate can be accounted for by the increased microviscosity of the solution. A comparison of the rate of degradation of DNA in an aqueous high temperature system with a formamide low temperature system showed that the reduced reaction rate was more than compensated for by a lower rate of degradation.

Schmeckpeper and Smith (29) found that the presence of 50% formamide reduces the rate of RNA-DNA hybridisation by a factor of 0.25 compared with the reaction rate in an aqueous solution at a similar stringency. The stability of RNA-DNA hybrids is greater in 50% formamide than that of DNA duplexes of similar base composition (22). This has been exploited for the purpose of electron microscopy to allow RNA-DNA hybrids to form whilst preventing DNA duplex formation. Using solution reactions, Vogelstein and Gillespie (31) and Casey and Davidson (32) showed that 70% formamide does not permit DNA reannealing at temperatures between 41 and 50°C but does allow almost complete RNA-DNA hybridisation.

6.6 Mismatching

It is well established that the presence of mismatched base pairs reduces the thermal stability of DNA duplex molecules. It is therefore to be expected that the imperfect sequence complementarity will also have an effect on the rate of reassociation. This is particularly important where reassociation is used to study the divergence of the DNAs of related species.

The extent of mismatching in a DNA duplex is experimentally determined by measuring the reduction in T_m of the mismatched DNA compared with a modified control DNA. Hence the effect of mismatching on the rate of DNA reannealing is usually quoted in terms of effective reduction in T_m. From several studies (20,27,33,34) it is clear that mismatching of DNA which results in a decrease in the T_m of about 15°C will reduce the reannealing rate by a factor of 2.

6.7 Temperature

The dependence of DNA reassociation rate on incubation temperature was first studied by Marmur and Doty (35). They observed that as the temperature was reduced from the T_m, the rate of reassociation increased until a maximum was reached at about 25°C below the T_m. Further reductions in the temperature reduced the reaction rate, resulting in the characteristic bell-shaped dependence of rate upon temperature. This observation has since been repeated many times by other workers. A similar dependence for RNA-DNA hybridisation has been obtained with both reactants free in solution (22,24), and with DNA bound to nitrocellulose filters (21). However, in contrast to DNA reassociation, the maximum rate of RNA-DNA hybridisation is obtained at about $10 - 15$°C below the T_m of the hybrids.

Strauss and Bonner (36) studied the temperature dependence of RNA-DNA hybridisation under conditions of moderate DNA excess. In this way the DNA-DNA reaction was allowed to compete with the RNA-DNA reaction. It was found that the relative rate of RNA-DNA hybridisation increased more rapidly with temperature than the equivalent DNA-DNA reaction. This result was explained by proposing that single-stranded RNA and DNA molecules form short base-paired intra-strand regions which suppress the rate by reducing the concentration of the available sites for nucleation.

6.8 Phenol Emulsion Technique

The presence of a phenol aqueous emulsion in a reaction has been shown to increase dramatically the rates of reassociation (37). This increase, however, was shown to be dependent on the concentration of the reacting DNA species. At low DNA concentration (4 µg/ml) the rate of reassociation was increased by a factor of 25 000, but at higher concentrations (1.6 mg/ml) the factor was only 35. The rate of RNA-DNA hybridisation in the phenol emulsion was also observed to be increased by up to 100-fold. These increases in rate are not fully understood, but it has been suggested that single-stranded DNA concentrates at the phenol-aqueous interface or forms aggregates or semi-precipitates elsewhere in the two-phase system. The rapid rate of reassociation thus may be due to high local DNA concentrations. The same principle is involved in the use of neutral or anionic dextran polymers to increase reaction rate (38). This phenol emulsion technique, which has potential for the reassociation of small amounts of genomic DNA, has not yet been widely used.

7. TECHNIQUES FOR MEASUREMENT OF REASSOCIATION

A major consideration in planning a reassociation experiment is the range of C_0t or R_0t over which the reaction is expected to take place. For example, *E. coli* DNA reassociates with a $C_0t_{1/2}$ of 3.5 mol. litre^{-1}. sec at 60°C in 0.12 M sodium phosphate buffer. Almost complete reassociation of *E. coli* DNA will therefore be achieved by a C_0t of 100 mol. litre^{-1}. sec [C_0t and R_0t values can be easily calculated from knowing that 83 µg/ml of nucleic acid incubated for 1 h is equivalent to a C_0t of 1 mol. litre^{-1}. sec]. The reassociation of the single-copy sequences in eukaryotic DNA requires incubation under standard conditions to C_0t values around 10^4 mol. litre^{-1} sec. The practical upper limit for the concentration of DNA in solution is about $5 - 10$ mg/ml and therefore time periods of the order of days may be required to ensure the reassociation of single copy sequences. Having calculated the approximate incubation time required, the next decision is the method of monitoring the reassociation. The reassociation of nucleic acid can be followed using a variety of techniques, each of which has advantages for particular applications. These are considered below.

7.1 **Optical Methods**

When double-stranded DNA is denatured there is an increase in the u.v. absorption (hyperchromicity) and as DNA is allowed to reanneal there is a corresponding decrease in the u.v. absorption (hypochromicity). It is therefore possible to quantitate DNA reannealing by following the change in the absorption at 260 nm with time. In general, such measurements are performed in a temperature-controlled cuvette placed in a spectrophotometer which provides a continuous read-out of u.v. absorption. The initial step is to raise the temperature to melt the DNA sample and to provide a maximum absorbance reading to be compared with that obtained at the end of the reaction. The sample temperature is then lowered as rapidly as possible to the incubation temperature and the change in u.v. absorption with time is recorded. The drop in temperature requires several minutes which, at a typical DNA concentration of around 50 μg/ml, is equivalent to a C_0t value of 0.01 mol. litre^{-1}. sec. It is therefore not practical to obtain information at earlier C_0t values. Since the repetitive DNA of higher organisms will have reassociated by such C_0t values, the optical technique is not appropriate for experiments involving the study of repetitive DNA.

7.2 **Hydroxyapatite**

The use of hydroxyapatite (HAP) for the analysis of DNA reassociation has been reviewed extensively by Britten *et al.* (17). At low phosphate ion concentrations (10 − 20 mM sodium phosphate buffer, pH 6.8) both single-stranded and double-stranded DNAs are adsorbed to HAP. However, at higher concentrations (0.12 − 0.14 M sodium phosphate buffer) single-stranded DNA is eluted and double-stranded DNA is retained. Double-stranded DNA can be eluted at a concentration of 0.4 − 0.5 M sodium phosphate buffer.

Elution is generally carried out in a water-jacketed column at 60°C thus maintaining the discrimination of the incubation reaction which is usually at the same temperature. It is possible to vary both the incubation temperature and the elution temperature in order to allow detection of mismatched sequences (39). The eluted DNA can be measured either by u.v. absorption or by scintillation counting if it is radioactively labelled. HAP has been used widely in experiments in which a radioactive tracer DNA is mixed with a large excess of genomic DNA (40). The reassociation of tracer DNA is measured by scintillation counting of eluted fractions and the reassociation of genomic DNA is measured optically from the same fractions. Such experiments have been used to measure gene copy number by comparing rates of reassociation.

Fractionation by HAP chromatography offers a number of advantages over other techniques for the measurement of DNA reassociation. It is simple to use and gives reproducible results. A wide range in the amount of DNA analysed is possible, with an upper limit of about 200 μg DNA/ml packed volume of HAP being advisable. If desired, the separated fractions can be recovered for further analysis. HAP is available commercially (e.g., BioRad) and each batch should be checked for its elution characteristics. It is important that a single batch is used for any given set of experiments and that an internal control is used when measuring reassociation kinetics.

7.3 **Nuclease S1**

The single strand-specific nuclease, S1, has been used to monitor both the reassociation of DNA and also RNA-DNA hybridisation. This technique measures the fraction of DNA nucleotides in duplexes at each point in the annealing reaction. For DNA reassociation, therefore, it should yield similar results to those obtained by the optical method described above. This contrasts with HAP chromatography which measures the fraction of DNA fragments that are totally single-stranded. A comparison of *E. coli* DNA reassociation measured by HAP and by nuclease S1 digestion showed that the time courses followed different kinetics (6). The HAP time course followed second-order kinetics, whereas a modified equation was required to describe the kinetics of the nuclease S1-derived time course (i.e., Equation 13). Nuclease S1 analysis may therefore not be appropriate for DNA reassociation; rather, HAP analysis may be the better technique for this purpose.

Nuclease S1 analysis has been used extensively for analysis of RNA-DNA annealing in RNA excess hybridisations. It has been shown that, under conditions of RNA excess, the reaction kinetics as measured by nuclease S1 digestion follow second-order reaction theory (3). Measurement of the reaction kinetics of complex mixtures of RNA species with their cDNA copies has allowed estimation of their complexity. Such measurements rely upon comparison with standardised reactions for a purified single mRNA species of known base sequence complexity. A practical example of such an experiment using nuclease S1 analysis is given in Section 7.4.

7.4 **Measurement of mRNA Complexities by $R_0 t$ Analysis**

In this worked example it is assumed that a complex mRNA mixture has been isolated and that it is desired to estimate the complexity of the population. It is also assumed that radioactively-labelled cDNA has been prepared using reverse transcriptase. The radio isotope used to label the DNA can be ^3H, ^{35}S or ^{32}P since detection is by liquid scintillation counting. The experiment consists of measuring the hybridisation kinetics of the mRNA with its cDNA using nuclease S1 and comparing the time course with that of a known standard.

7.4.1 *Experimental Strategy*

This procedure involves setting up a series of small reaction volumes in sealed capillaries each containing the same concentration of RNA and a small amount of labelled cDNA. A range of $R_0 t$ is achieved by incubating the capillaries for different periods of time from a few minutes to a few days. If an approximate value for the $R_0 t_{1/2}$ is known then the RNA concentration can be chosen to ensure that a range of points spanning this region will be obtained. If the $R_0 t_{1/2}$ is totally unknown then several different series of hybridisations each at different RNA concentrations will be necessary. It is desirable that the different series of points span over-lapping ranges of $R_0 t$ to ensure that no artificial discontinuities are introduced into the curve. In planning a $R_0 t$ curve it is useful to remember than an RNA concentration of 83 μg/ml incubated for 1 h is equivalent to a $R_0 t$ value of 1.0 mol. litre^{-1}. sec. In the present example it is necessary to span a range of $R_0 t$ of $10^{-3} - 10^2$ mol. litre^{-1}. sec and therefore it is necessary to set up two series of reactions at 5 μg RNA/ml (series A) and 500 μg RNA/ml (series B). The following incubation times are suggested:

	$R_o t$ (mol. litre^{-1}. sec)	
Time	*Series A*	*Series B*
0	0	0
1 min	0.001	0.1
2 min	0.002	0.2
4 min	0.004	0.4
5 min	0.005	0.5
10 min	0.010	1.0
15 min	0.015	1.5
20 min	0.02	2.0
30 min	0.03	3.0
40 min	0.04	4.0
1 h	0.06	6.0
2 h	0.12	12.0
3 h	0.18	18.0
4 h	0.24	24.0
6 h	0.36	36.0
8 h	0.48	48.0
10 h	0.60	60.0
16 h	0.96	96.0
24 h	1.44	144.0
30 h	1.80	180.0

After incubation, the amount of cDNA in hybrids is measured by digestion of unhybridised cDNA with nuclease S1 followed by precipitation with trichloroacetic acid (TCA). The total amount of cDNA in each hybridisation reaction is also measured by precipitation with TCA without nuclease S1 digestion. The results of series A and series B hybridisations are combined and the proportion of cDNA in hybrid is plotted as a function of log ($R_o t$).

7.4.2 *Reagents Required*

(i)   10 x hybridisation buffer (3 M NaCl, 10 mM EDTA, 1 M Tris-HCl, pH 7.5), prepared using sterile distilled water. The pH is adjusted by adding NaOH to the 10 x buffer and checking the pH of 10-fold diluted aliquots at 70°C.

(ii)   Poly(U) (5 mg/ml).

(iii)   Nuclease S1 buffer: 2.8 mM $ZnSO_4$

0.14 M NaCl

70 mM sodium acetate, pH 4.5

14 μg/ml heat-denatured, sonicated salmon sperm DNA

(see Chapter 5, *Table 3,* footnote b)

(iv)   Nuclease S1 (Sigma).

(v)   Sonicated salmon sperm DNA (1 mg/ml) as carrier.

(vi)   10% TCA and 5% TCA.

7.4.3 *Procedure*

(i)   Two series of reactions are required — series A (5 µg RNA/ml) and series B
        (500 µg RNA/ml). Thus, series A consists of 20 capillaries each containing 50 ng
        of RNA and 5000 c.p.m. of cDNA in 10 µl. Series B consists of 20 capillaries
        each containing 500 ng of RNA and 5000 c.p.m. of cDNA in 1 µl. To do this,
        mix the following quantities:

	Series A	Series B
mRNA (µg)	1	10
cDNA (c.p.m.)	100 000	100 000
poly(U) (µg)	100	—
10 x hybridisation buffer (µl)	20	2
Final volume	200 µl	20 µl

Poly(U) is added to series A reactions to make the total RNA concentration equal to
that of series B. If the volume is greater than the intended final volume, lyophilise the
sample to dryness and then make up to the final volume indicated with sterile distilled
water.

(ii)   Dispense 10 µl aliquots (series A) and 1 µl aliquots (series B) into sterile siliconised
        capillaries and heat-seal both ends using a small bunsen burner. The capillaries
        can now be stored at −20°C until ready for use.

(iii)   Heat each capillary at 100°C for 3 min to denature the nucleic acid contents
        and incubate at 70°C for the times indicated in Section 7.4.1. All incubations
        are done by completely immersing the capillaries in a water bath.

(iv)   After incubation, open the capillaries with a glass knife and flush out the con-
        tents of each with 210 µl of nuclease S1 buffer into a numbered plastic microcen-
        trifuge tube. At this stage, each sample can be stored at −20°C until ready for
        analysis if desired.

(v)   Divide each sample accurately into two equal aliquots of 100 µl, appropriately
        numbered and marked P or T.

(vi)   Add to one of each pair, (P), approximately 30 units of nuclease S1 and incubate
        for 2 h at 37°C. The activity of the nuclease should be checked prior to this
        analysis using unhybridised cDNA. Also incubate the other member of each pair
        (T) but without added nuclease.

(vii)   Add 100 µl of carrier DNA to each aliquot (both P and T series). Mix and then
        add 2 ml of 10% TCA to each aliquot and keep on ice for 15 min.

(viii) Filter each aliquot onto a Whatman GF/C filter using a vacuum filter device
        and wash each carefully three times with 5% TCA.

(ix)   Wash each filter twice with ethanol.

(x)   Place each filter in a numbered counting vial and heat at 60°C for 1 h to remove
        ethanol.

(xi)   Add 10 ml scintillation fluid to each vial and measure the radioactivity using
        a liquid scintillation counter.

(xii)   Plot the radioactivity as percentage cDNA hybridised (i.e., P/T x 100) *versus*
        log ($R_o t$).

7.4.4 *Analysis of Data*

As described earlier (Section 4.2), the analysis of the reaction kinetics of a complex mRNA population with its cDNA requires that the time course is analysed into a number of components. Typical data which were obtained by the above protocol are shown in *Figure 4b* and their analysis into three kinetic components using the least squares fitting program (Appendix III) is shown in *Table 2*. The kinetic standard used was globin mRNA which has a molecular weight of 4×10^5 daltons and a $R_0 t_{1/2}$ of 6.6×10^{-4} mol. litre^{-1}. sec.

8. REFERENCES

1. Wetmur,J.G. and Davidson,N. (1968) *J. Mol. Biol.*, **31**, 349.
2. Britten,R.J. and Kohne,D.E. (1968) *Science (Wash.)*, **161**, 529.
3. Galau,G.A., Britten,R.J. and Davidson,E.H. (1977) *Proc. Natl. Acad. Sci. USA*, **74**, 1020.
4. Galau,G.A., Smith,M.J., Britten,R.J. and Davidson,E.H. (1977) *Proc. Natl. Acad. Sci. USA*, **74**, 2306.
5. Melli,M., Whitfield,C., Rao,K.V., Richardson,M. and Bishop,J. (1971) *Nature New Biol.*, **231**, 8.
6. Smith,M.J., Britten,R.J. and Davidson,E.H. (1975) *Proc. Natl. Acad. Sci. USA*, **72**, 4805.
7. Britten,R.J. and Davidson,E.H. (1976) *Proc. Natl. Acad. Sci. USA*, **73**, 415.
8. Young,B.D. and Paul,J. (1973) *Biochem. J.*, **135**, 573.
9. Marquadt,D.W. (1963) *J. Soc. Ind. Appl. Math.*, **11**, 431.
10. Bevington,P.R. (1969) *Data Reduction and Error Analysis for the Physical Sciences*, McGraw-Hill Book Co., New York.
11. Pearson,W.R., Davidson,E.H. and Britten,R.J. (1977) *Nucleic Acids Res.*, **4**, 1727.
12. Kells,D.J.C. and Strauss,N.A. (1977) *Anal. Biochem.*, **80**, 344.
13. Hell,A., Young,B.D. and Birnie,G.D. (1976) *Biochim. Biophys. Acta*, **442**, 37.
14. Bishop,J.O., Morton,J.G., Rosbash,M. and Richardson,M. (1974) *Nature*, **250**, 199.
15. Young,B.D., Harrison,P.R., Gilmour,R.S., Birnie,G.D., Hell,A., Humphries,S. and Paul,J. (1974) *J. Mol. Biol.*, **84**, 555.
16. Kennell,D.E. (1971) *Prog. Nucleic Acid Res. Mol. Biol.*, **11**, 259.
17. Britten,R.J., Graham,D.E. and Neufeld,B.R. (1974) in Grossman,L. and Moldave,K. (eds.), *Methods in Enzymology*, Vol. **29**, Academic Press, London and New York, p. 363.
18. Wetmur,J.G. (1976) *Ann. Rev. Biophys. Bioeng.*, **29E**, 363.
19. Wetmur,J.G. (1971) *Biopolymers*, **10**, 601.
20. Hutton,J.R. and Wetmur,J.G. (1973) *J. Mol. Biol.*, **77**, 495.
21. Birnsteil,M., Sells,B. and Purdom,I.F. (1972) *J. Mol. Biol.*, **63**, 21.
22. Bishop,J.O. (1972) *Biochem. J.*, **126**, 171.
23. Studier,F.W. (1969) *J. Mol. Biol.*, **41**, 199.
24. Nygaard,A.P. and Hall,B.D. (1964) *J. Mol. Biol.*, **9**, 125.
25. Thrower,K.J. and Peacocke,A.R. (1968) *Biochem. J.*, **109**, 543.
26. Subirana,J.A. and Doty,P. (1966) *Biopolymers*, **4**, 171.
27. Chang,C.T., Hain,T.C., Hutton,J.R. and Wetmur,J.G. (1974) *Biopolymers*, **13**, 1847.
28. McConaughy,B.L., Laird,C.D. and McCarthy,B.J. (1969) *Biochemistry (Wash.)*, **8**, 3289.
29. Schmeckpeper,B.J. and Smith,K.D. (1972) *Biochemistry (Wash.)*, **11**, 1319.
30. Hutton,J.R. (1977) *Nucleic Acids Res.*, **4**, 3537.
31. Vogelstein,B. and Gillespie,D. (1977) *Biochem. Biophys. Res. Commun.*, **75**, 1127.
32. Casey,J. and Davidson,N. (1977) *Nucleic Acids Res.*, **4**, 1539.
33. Lee,C.H. and Wetmur,J.G. (1973) *Biochem. Biophys. Res. Commun.*, **50**, 879.
34. Miller,S.J. and Wetmur,J.G. (1974) *Biopolymers*, **13**, 2545.
35. Marmur,J. and Doty,P. (1961) *J. Mol. Biol.*, **3**, 585.
36. Strauss,N.A. and Bonner,T.I. (1972) *Biochim. Biophys. Acta*, **277**, 87.
37. Kohne,D.E., Levison,S.A. and Byers,M.J. (1977) *Biochemistry (Wash.)*, **16**, 5329.
38. Wetmur,R. (1975) *Biopolymers*, **14**, 2517.
39. Young,B.D. and Paul,J. (1975) *J. Mol. Biol.*, **96**, 783.
40. Harrison,P.R., Birnie,G.D., Hell,A., Humphries,S., Young,B.D. and Paul,J. (1974) *J. Mol. Biol.*, **84**, 539.

CHAPTER 4

Quantitative Filter Hybridisation

MARGARET L.M. ANDERSON and BRYAN D. YOUNG

1. INTRODUCTION

An application of nucleic acid hybridisation, which is of central importance to genetic engineering and is finding increasing use in molecular biology, is filter hybridisation. The technique is derived from the classical experiments of Gillespie and Speigelman (1). Denatured DNA or RNA is immobilised on an inert support, for example nitrocellulose, in such a way that self-annealing is prevented, yet bound sequences are available for hybridisation with an added nucleic acid probe. To facilitate analysis, the probe is labelled, often with ^{32}P. Hybridisation is followed by extensive washing of the filter to remove unreacted probe. Detection of hybrids is usually by autoradiography although, when the hybrids are sufficiently radioactive, scintillation counting can be used. The procedure is widely applicable, being used for phage plaque and bacterial colony hybridisation, Southern and Northern blot hybridisation, dot blot hybridisation and hybrid selection (see other chapters in this volume).

While solution hybridisation is the standard method for quantitative measurements of sequence complexity and composition (2), there are practical difficulties when the number of samples is large. By contrast, dot blot hybridisation is ideally suited to the analysis of multiple samples. The technique has the added advantage that it is easy to prepare replicate filters allowing many filter-bound sequences to be analysed at the same time, for example with different probes or under different hybridisation and washing conditions. Dot blot hybridisation can be used qualitatively since it is capable of great discrimination, as exemplified by the ability to distinguish between closely similar members of multigene families (3,4). It can also be used quantitatively with appropriate calibration (5), but it is most commonly used as a semi-quantitative method for determining the relative levels of sequences in different samples. As we shall see, its use is limited by the low rate of hybridisation and by a level of sensitivity which makes it less useful than solution hybridisation for analysing rare sequences.

In this chapter we will first discuss theoretical aspects of filter hybridisation (Sections 2 − 5) and then describe practical aspects (Sections 6 − 14). To make clear the distinction between sequences which are in solution and those which are filter-bound, we will use a different nomenclature from that in the previous chapter. Subscripts 's' and 'f' will be used for nucleic acid in solution and filter-bound nucleic acid, respectively.

2. KINETICS OF FILTER HYBRIDISATION

Nucleic acid hybridisation depends on the random collision of two complementary sequences. As described in the previous chapter, the time course of the reaction in solution is determined by the concentration of the reacting species and by the second order rate constant, k. The stability of the duplex formed is dependent on its melting temperature, T_m. For hybridisation of perfectly-matched complementary sequences in solution, equations have been derived which describe the process fairly precisely (e.g., Chapter 3, Equations 4, 5 and 10). Detailed investigations have determined the effects of changes in reaction conditions for solution hybridisation, so that values for k and T_m can be calculated with some confidence. In contrast, filter hybridisation has been less extensively studied, and the parameters affecting the rate and extent of reaction are less well understood. Calculations made from solution hybridisation are not necessarily valid for filter hybridisation, although changes in reaction conditions probably have a similar qualitative effect.

The hybridisation of a denatured nucleic acid probe in solution to a filter-bound nucleic acid is a function of two competing reactions, *viz*. the reassociation of sequences in solution and the hybridisation to filter-bound DNA or RNA. (Since the filter-bound nucleic acid is immobilised, reassociation of bound sequences does not occur.) The rate of disappearance of single strands may be expressed by the equation:

$$-\frac{d[C_s]}{dt} = k_1[C_f][C_s] + k_2[C_s]^2 \qquad \text{Equation 1}$$

where C_f is the concentration of filter-bound nucleic acid sequence, C_s is the concentration of nucleic acid probe in solution, k_1 is the rate constant for the hybridisation reaction on the filter and k_2 is the rate constant for the reassociation in solution. The term $k_1[C_f][C_s]$ represents the filter hybridisation while the term $k_2[C_s]^2$ represents reassociation in solution.

A variety of factors affect the rate constants (see Section 3). The rate constant should be the same for both reactions (i.e., $k_1 = k_2$) provided that:

(i) the nucleation rates at the filter and in solution are the same
(ii) the effective molecular weight of the nucleic acid species in solution is smaller than that of the filter-bound nucleic acid, since the rate constant for DNA-DNA reassociation is dependent on the size of the smaller fragment (6,7), at least in solution.

Equation 1 predicts that the initial rate of hybridisation is proportional to the concentrations of both the probe in solution and the filter-bound sequences. When $[C_f]$ is much higher than $[C_s]$, as in dot blots of plasmid DNA (this Chapter, Section 6.2.1 and Chapter 5, Section 3.1) or in hybrid selection (Chapter 5, Section 3.2), the solution reassociation term can be ignored and Equation 1 simplifies to the pseudo-first order reaction:

$$-\frac{d[C_s]}{dt} = k_1[C_f][C_s] \qquad \text{Equation 2}$$

where $[C_f]$ is constant.

On integration, this gives:

$$\frac{[C_s]_t}{[C_s]_0} = e^{-k_1[C_f]t}$$

Equation 3

where $[C_s]_t$ is the value of $[C_s]$ at time t.

While it has been shown experimentally that the initial hybridisation rate is proportional to $[C_s]$, the relationships in Equations 1 and 2 do not describe exactly the dependency of the hybridisation rate on $[C_f]$. At low values of $[C_f]$, the initial rate of hybridisation is proportional to $[C_f]$, but the rate does not increase linearly at higher values (8 – 10). This is explained by the fact that filter hybridisation depends on two processes, diffusion of the probe to the filter and hybridisation at the filter. It is thought that at low values of $[C_f]$, the hybridisation reaction itself is the rate-limiting step, whereas at high values of $[C_f]$ hybridisation is so fast that the solution surrounding the filter becomes depleted of probe and the overall reaction is then limited by diffusion of the probe to the filter. Flavell *et al.* (9) have shown that at higher values of $[C_f]$ the rate equation should incorporate a term, J, to take diffusion of the probe into account. Therefore, at high values of $[C_f]$, Equation 1 can be replaced by an equation of the form:

$$\frac{-d[C_s]}{dt} = J + k_2[C_s]^2$$

Equation 4

where $k_1 [C_f] [C_s] > J > 0$. The diffusion term J is a function of the diffusion coefficient of the probe and the concentration gradient of the probe. The relationship between J and $k^2[C_s]^2$ determines whether reassociation of the probe is an important factor. When J is $\leq k_2[C_s]^2$, reassociation will be significant.

Since many filter hybridisation experiments aim to hybridise the maximum amount of probe to excess sequences on the filter, the relationship given in Equation 4 is important. The overall hybridisation reaction will be speeded up by factors which increase diffusion of the probe to the filter, for example, using a small probe, high incubation temperature, low reaction volume and shaking the reaction vessel. The rate constant k_1 can be determined by two methods.

(i)   From initial reaction rates. Rearranging the terms in Equation 2,

$$k_1 = \frac{v_i}{[C_f][C_s]}$$

where v_i is the initial rate of reaction. So in a filter-bound DNA excess reaction, a plot of the reciprocal of the percentage of added probe which has hybridised to the filter *versus* the reciprocal of the time of reaction will give a straight line with a slope of $1/k_1$ (Chapter 3, Section 3.3.1). This holds true for both double-stranded and single-stranded probes, although a small correction may have to be made in the value of $[C_s]$ for reassociation of the probe in solution.

(ii)   From measuring $t_{1/2}$, that is, the time when $[C_s]_t/[C_s]_0$ is 0.5. For a filter-bound DNA-excess reaction (pseudo-first order kinetics) a plot of log [fraction of the probe remaining single-stranded] against the time of reaction will give a straight line. The $t_{1/2}$ can be read off the graph and substituted in Equation 3 to give:

$$[C_f]t_{1/2} = \frac{\ln2}{k_1} \qquad \text{Equation 5}$$

or

$$k_1 = \frac{0.693}{[C_f]t_{1/2}} \qquad \text{Equation 6}$$

Experimentally, values for k_1 obtained from $t_{1/2}$ measurements and from initial rate data for hybridisation of a simple DNA probe to filter-bound DNA are in good agreement. However, the values obtained are 10 times lower than those obtained for solution hybridisation of the same DNAs (9). The reason may be that only a fraction of the DNA bound to the filter is accessible for nucleation, although all the DNA can effectively participate in hybrid formation. Thus the concentration term $[C_f]$ used to calculate the rate constant may be incorrect and k_1 (hybridisation) may actually be equal to k_2 (reassociation). Alternatively, rate constants may be lower for filter hybridisation. As a consequence of binding nucleic acid sequences to the filter, steric restraints may retard the formation of stable nucleation complexes.

Equations 1 and 4 show that one of the factors affecting the kinetics of a filter hybridisation reaction is reassociation of the probe. This variable is often overlooked but its effects can be large and can cause problems in interpreting results. It has been shown that as much as $20-30\%$ of the input DNA probe can be unavailable for hybridisation due to reassociation (9). A second complication is that the probe may form concatenates of partially-reassociated duplexes with single-stranded regions which can hybridise to filter-bound sequences. Again the effects are not negligible. Flavell *et al.* (11) showed that 10% of the added denatured, double-stranded DNA which hybridised to a filter containing single-stranded DNA represented homologous rather than complementary sequences. Similar problems can arise in DNA-RNA hybridisation experiments with self-complementary transcripts. In order to minimise these complications, it is desirable to choose reaction conditions which facilitate diffusion of probe to the filter and favour hybridisation over reassociation, that is, use of a small probe (preferably single-stranded), small reaction volume, a low concentration of probe in solution and a high reaction temperature.

3. FACTORS AFFECTING THE RATE OF FILTER HYBRIDISATION

3.1 Concentration of the Probe

There has been no systematic study of effects of the concentration of probe on the rate of hybridisation at the filter and on the yield of duplex. However, the following points should be noted.

3.1.1 *Double-stranded Probe in Excess*

If the probe is a simple double-stranded sequence, Equation 1 predicts that at high $[C_s]$ values, reassociation of the probe should be favoured over hybridisation to the bound nucleic acid. Therefore, as incubation continues, the reaction will change from being in probe excess to being in filter-bound excess where, as we have seen, the kinetics are different. Increasing the concentration of probe in solution, $[C_s]$, will increase the initial rate of hybridisation at the filter and the proportion of filter-bound sequences in duplex will increase, but not dramatically. For DNA probes and filter-bound RNA, as in RNA dot blots, high concentrations of formamide can be used to suppress reassociation in solution (see Section 4.1.2).

3.1.2 *Single-stranded Probe*
Whether in excess or not, there is no reassociation of a single-stranded probe in solution unless there are regions of extensive self-complementarity. The rate of hybridisation to the filter and the amount of hybrid formed should increase with increase in $[C_s]$. It is important to note, however, that the probe concentration should not be increased without limit. If more than about 100 ng ^{32}P-labelled probe per ml is used, non-specific irreversible binding to the filter occurs.

3.2 **Probe Complexity**

For solution hybridisation, the rate of reassociation of DNA is an inverse function of its complexity, so that the more complex the DNA, the slower the rate of reassociation (2,12). Extending this to filter hybridisations, the rate of reassociation of the probe should fall when the complexity of the DNA increases and its effective $[C_s]$ decreases. This is indeed what is observed (9). In contrast, two effects of complexity are seen for hybridisation of the probe to filter-bound nucleic acid sequences. When $[C_f]$ is low, the rate of hybridisation is inversely proportional to complexity over a 400-fold range, indicating that the reaction is controlled by the nucleation step. However, when the hybridisation reaction is limited by diffusion of the probe to the filter, that is when $[C_f]$ is high, the rate of reaction is independent of complexity (9).

3.3 **Molecular Weight of the Probe**

For DNA-DNA hybridisation in solution, the rate is directly proportional to the square root of the molecular weight of the nucleic acid (12) and this also describes the reassociation of the probe in solution during filter hybridisation (9). However, the effect of the molecular weight of the probe on the rate of hybridisation to filter-bound sequences contrasts sharply with that found in solution. Two situations can occur. When $[C_f]$ is low compared with $[C_s]$, that is, a nucleation-limited reaction, the rate of hybridisation is independent of the molecular weight (9,10). When $[C_f]$ is high compared with $[C_s]$, that is, diffusion-limited filter hybridisation, the rate of hybridisation is inversely proportional to the molecular weight of the probe, but there are insufficient data for an exact relationship to be formulated. The observed rate of hybridisation is significantly depressed by an increase in the molecular weight of a single-stranded probe (which is not capable of reassociation). This effect is even more pronounced when a double-stranded probe is used. This is because the combined effects of a lower rate of hybridisation and the increased rate of reassociation, which accompanies an increase in molecular weight of the sequences in solution, result in lower observed rates of hybridisation and a reduced final yield of hybrid. The difference in dependence on molecular weight of the two types of filter hybridisation is not understood.

3.4 **Base Composition**

The base composition of nucleic acids affects the rate of hybridisation, the rate increasing with increasing % G+C. However, the effect is small (12) and can be ignored in practice.

3.5 **Temperature**

The temperature of reaction affects the rate of any hybridisation reaction (13). Typically a bell-shaped temperature dependence curve is obtained. At 0°C, hybridisation proceeds extremely slowly, but as the temperature is raised, the rate increases dramatically to reach a broad maximum which is $20-25$°C below T_m for DNA-DNA annealing. At higher temperatures the duplex molecules tend to dissociate so that as the temperature approaches T_m -5°C, the rate is extremely low. The relationship applies to the formation of both well-matched and poorly-matched hybrids although the curve is displaced towards lower temperatures for mismatched duplexes (14). So, ideally, hybridisations should be carried out at a T_i (incubation temperature) that is $20-25$°C below T_m. In practice, for well-matched hybrids, the hybridisation reaction is usually carried out at 68°C in aqueous solution and at 42°C for solutions containing 50% formamide. For poorly-matched hybrids, incubation is generally at $35 - 42$°C in formamide-containing solutions.

A similar dependence has been shown for RNA-DNA hybridisations (10), but here the maximal rate of hybridisation is obtained at some $10-15$°C below the T_m of the hybrids.

3.6 **Formamide**

Formamide decreases the T_m of nucleic acid hybrids (see Section 4.1). This is a very useful property because by including $30-50$% formamide in the hybridisation solution, the incubation temperature, T_i, can be reduced to $30-42$°C. This has several practical advantages: the probe is more stable at lower temperatures, there is better retention of non-covalently-bound nucleic acid on the filter and nitrocellulose filters are less likely to disintegrate at the lower temperature.

Concentrations of formamide between 30 and 50% apparently have no effect on the rate of filter hybridisation and 20% formamide reduces the rate by only about one-third (15). On the other hand, a concentration of 80% formamide is thought to depress the rate constant for hybridisation in solution at least by a factor of three for DNA-DNA duplexes and by a factor of 12 for RNA-DNA hybrids (16). Qualitatively similar results are likely to occur in filter hybridisation.

Formamide can be used to alter the stringency of the reaction conditions. By holding T_i constant and varying the concentration of formamide, different effective temperatures are obtained. Effective temperatures as low as 50°C below the T_m of perfectly-matched hybrids can be reached which allows detection of homologies with as much as 35% mismatching (15).

3.7 **Ionic Strength**

At low ionic strength, nucleic acids hybridise very slowly, but as the ionic strength increases, the reaction rate increases. The effect is most dramatic at low salt concentrations (<0.1 M Na$^+$) where a 2-fold increase in concentration increases the rate 5- to 10-fold. Above 0.1 M Na$^+$ the rate dependence is less, but still marked up to about 1.5 M Na$^+$ (12,17).

High salt concentrations stabilise mismatched duplexes, so to detect cross-hybridising species, the salt concentration of hybridisation and washing solutions must be kept fairly high. Washing is therefore generally carried out using 2 − 6 x SSC (1 x SSC is 0.15 M NaCl, 0.015 M trisodium citrate, pH 7.0).

3.8 Dextran Sulphate

Wetmur (18) observed that the addition of an inert polymer such as dextran sulphate increased the rate of hybridisation in solution. Thus the presence of 10% dextran sulphate gave rise to a 10-fold increase in reassociation rate. The effect was attributed to the exclusion of the DNA from the volume occupied by the polymer, that is, the dextran sulphate effectively increased the concentration of the DNA. A qualitatively similar effect occurs in filter hybridisation using both DNA and RNA probes (19) where, of course, the concentrating effect of the polymer applies only to the solution phase. For a single-stranded probe, the rate of hybridisation increases by 3- to 4-fold. For a double-stranded probe, the rate apparently increases by up to 100-fold and the yield of hybrid apparently also increases. However, in both cases, most of this increase is caused by the formation of concatenates which readily occurs under these conditions, that is, extensive networks of reassociated probe which, by virtue of single-stranded regions, hybridise to filter-bound nucleic acid and so lead to over-estimation of the extent of hybridisation. For qualitative studies this amplification in the hybridisation signal caused by binding of labelled probe is quite useful. However, for quantitative studies the effect may complicate the interpretation of results. Therefore it may be desirable to reduce the likelihood of networks forming by using probes which are not self-complementary. If double-stranded DNA probes are used they should be short to minimise the formation of extensive networks of probe. For example, if nick-translated, double-stranded DNA probes (Chapter 2, Section 4.1.2) are used, the DNase concentration in the nick-translation reaction should be adjusted to give fragments ≤ 400 nucleotides long. Short incubation times should also be used since the formation of networks occurs late in the reaction. Finally it should be noted that solutions of dextran sulphate are viscous (and so can be difficult to handle) and can lead to high backgrounds.

3.9 Mismatching

Many hybridisation reactions involve complex mixtures of sequences and the duplexes formed are not all perfectly base-paired. Mismatching has the effect of lowering the rate of hybridisation and the melting temperature of hybrids, T_m. The temperature dependence of k_i, the rate constant for the formation of mismatched hybrids, still gives a bell-shaped curve (Section 3.5), but k_i is lower and reaches its optimum at a lower temperature relative to the rate constant for formation of perfect hybrids (4,14). This has not been studied extensively, but available data suggest that if the reaction is carried out at a temperature which is optimal for the formation of mismatched sequences, that is, about 25°C below their T_m, the rate is reduced by a factor of two for every 10% mismatch (14).

3.10 Viscosity

As the viscosity of the solution increases, the rate of hybridisation decreases. The effect can be quite large, but there is insufficient data to formulate an exact relationship.

3.11 pH

The effect of pH has not been studied extensively but within the pH range 5−9, the rate of hybridisation at 0.4 M Na$^+$ is essentially independent of pH (ref. 12). In practice, hybridisation experiments are usually carried out at pH 6.8−pH 7.4.

4. FACTORS AFFECTING HYBRID STABILITY

The melting temperature T_m is a measure of the thermal stability of hybrids. No systematic study of the effect of different parameters has been made for filter hybridisation. However, in general, variables that alter the rate constant, k, also alter the T_m in the same direction. The relationships below are derived from studies on hybridisation in solution, but are expected to be similar, qualitatively at least, for filter hybridisation. It is worth noting, however, that as a consequence of binding nucleic acid to the filters, the T_m of hybrids is often lower than would be predicted from solution hybridisation studies (5).

4.1 Perfectly-matched Hybrids

4.1.1 *DNA-DNA Hybrids*

Many studies on the stability of perfectly-matched DNA duplexes in solution have shown that T_m is dependent on ionic strength, base composition and denaturing agents (14, 20,21). The following relationship has been derived from combining several results (15):

$$T_m = 81.5 + 16.6 \ (\log M) + 0.41 \ (\%G+C) \ -0.72 \ (\% \ \text{formamide}) \qquad \text{Equation 7}$$

where M is the molarity of the monovalent cation and (% G+C) is the percentage of guanine and cytosine residues in the DNA. The monovalent cation dependence holds between the limits of 0.01 −0.4 M NaCl, but only approximately above this (20). The T_m is maximal at 1.0−2.0 M NaCl. The dependence on (% G+C) is valid between 30% and 75% (G+C) (ref. 22). The reduction in T_m by formamide is greater for poly(dA:dT) (0.75°C per 1% formamide) than for poly(dG:dC) (0.5°C per 1% formamide) (ref. 16).

In aqueous solution at 1 M NaCl (equivalent to 6 x SSC), Equation 7 simplifies to:

$$T_m = 81.5 + 0.41 \ (\% \ G+C)$$

The following relationships, derived from solution hybridisation studies, are also useful:

(i)   Every 1% mismatching of bases in a DNA duplex reduces the T_m by 1°C (ref. 14).

(ii) $(T_m) \ \mu_2 - (T_m) \ \mu_1 = 18.5 \ \log \mu_2/\mu_1$
       where μ_1 and μ_2 are the ionic strengths of the two solutions (ref. 23).

4.1.2 *RNA-DNA Hybrids*

For RNA-DNA hybrids, the term in Equation 7 incorporating formamide concentration does not hold because the relationship between formamide concentration and the depression of T_m is not linear. At 80% formamide, RNA-DNA hybrids are more stable than DNA-DNA hybrids by some $10-30°C$ depending on the sequence (5,16). Carrying out the reaction in 80% formamide can therefore also be used to suppress formation of DNA-DNA duplexes and preferentially select RNA-DNA hybrids (5,16,24).

4.2 **Mismatched Hybrids**

The T_m of nucleic acid hybrids is depressed by base mismatching. Values obtained from solution hybridisation studies show that a 1% mismatch reduces the T_m by between 0.5 and 1.4°C (refs. 17,21,22,25,26). The exact figure depends on the (G+C) content of the DNA. The stability of the hybrids also depends on the distribution of mismatched bases in the duplex. Thus if two sequences have 20% base pair mismatch, the hybrid formed between them will have a high T_m if the mismatch is concentrated in one region leaving a long stretch of perfectly-matched duplex. In contrast, the hybrid will be extremely unstable if every fifth base is mismatched.

At high concentrations of salt, mismatched hybrids are more stable than at low concentrations. In practice this is very useful because varying the salt concentration can be used to stabilise or dissociate mismatched hybrids according to the requirements of the experiment.

5. DISCRIMINATION BETWEEN RELATED SEQUENCES

5.1 **Stringency of hybridisation**

A sizeable fraction of the eukaryotic genome is composed of families of similar, but not identical, sequences. It is often the aim of filter hybridisation studies to distinguish between closely- and distantly-related members of such a family, for example, in screening recombinant libraries or determining gene copy numbers by Southern blots (Chapter 5). In practice this means that reaction conditions must be adjusted to optimise hybridisation of one species and minimise hybridisation of others.

As explained in Section 3.5, bell-shaped curves describe the relationship between the rate of hybridisation and the temperature of incubation for formation of both well-matched and poorly-matched hybrids. For a poorly-matched hybrid, the rate constant is lower and the curve is displaced towards lower temperatures. When the ratio of rate constants (discrimination ratio) for cross-hybridisation and for self-hybridisation is plotted against temperature of reaction, a sigmoidal curve is obtained (*Figure 1*). At low temperatures, the ratio is high while at higher temperatures (approaching $T_m -20°C$ for perfectly-matched hybrids), the ratio approaches zero (4,14). Although the data are not extensive, Beltz *et al.* (4) have suggested that this curve is probably a member of a family of sigmoidal curves whose exact dependence on temperature depends on the degree of mismatching of the hybrids. The relationship is useful in that it predicts that it should be easier to distinguish between distantly-related sequences by incubating at low temperatures while it should be easier to distinguish closely-related sequences by hybridising at high temperatures.

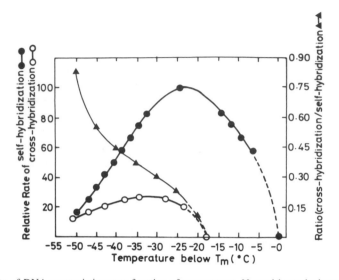

Figure 1. Rate of DNA reassociation as a function of temperature. Normal bacteriophage T4 DNA was used to examine the reassociation of perfectly-matched sequences (self-hybridisation; ●—●) and T4 DNA partially deaminated with nitrous acid was used for mismatched sequences (cross-hybridisation; ○—○). The dotted lines are extrapolations assuming that the rates of reassociation are zero at the appropriate T_m; for normal T4 DNA under these conditions (0.15 M sodium phosphate buffer) the T_m is 81°C. The discrimination ratio (▲—▲) is the rate constant for cross-hybridisation k_i, divided by the rate constant for self-hybridisation, k. Reproduced from reference 4 with permission.

In practice, therefore, to distinguish between the distantly-related members of a family of sequences, hybridisation should take place at a very permissive (relaxed) criterion. To detect closely-related members, the hybridisation should be at a stringent criterion. A single compromise criterion will not be effective because, as we have seen, different members of the family probably have different discrimination *versus* temperature curves. Hybridisation at a relaxed criterion followed by washing under progressively more stringent conditions may be useful for detecting distantly-related members of a family, but is not suitable for identifying closely-related members. This is probably because hybridisation and washing depend on different parameters. Hybridisation depends on the nucleation frequency while washing depends on the thermal stability (T_m) of the hybrids. Thus, a stringent hybridisation followed by a stringent wash is better for detecting closely-related members of a family than permissive hybridisation and a stringent wash.

5.2 **Extent of Reaction**

In distinguishing between related sequences, it is important to consider the extent of reaction. At first sight, it might appear that the longer the time of incubation the better should be the discrimination, but this is not the case. The following arguments have been made by Beltz *et al.* (4).

When two (or more) filter-bound sequences react with the same probe, the rate of depletion of the probe is given by the following equation:

$$\frac{-dC_s}{dt} = k_1[C_f][C_s] + k_2[C_s]^2 + k_i[C_i][C_s] \qquad \text{Equation 8}$$

where $[C_s]$ here refers to the concentration of the probe in solution at time zero.

This equation is derived from Equation 1 by the addition of a term to allow for cross-reaction of the probe with a related, filter-bound sequence, i, which has a concentration $[C_i]$ and a hybridisation rate constant k_i. $[C_f]$ is the concentration of filter-bound sequence, f, which is identical to the probe. The kinetics differ considerably depending upon which sequences are in excess and whether the probe can reassociate. For simplicity, in the following analyses (Sections 5.2.1 − 5.2.3) we have assumed that the concentrations of filter-bound sequences are the same (that is $[C_f] = [C_i]$) and $k_1 = k_2$. In fact we know that the rate constants are not equal, but the result will be qualitatively the same.

5.2.1 *Filter-bound Nucleic Acid in Excess*

When the filter-bound sequences are in excess over the probe, as in typical plasmid DNA dot blots (Section 6.2.1), Equation 8 simplifies to a pseudo-first order reaction where the rate of loss of the probe ($-dC_s/dt$) is given by:

$$\frac{-dC_s}{dt} = k_1[C_f][C_s] + k_i[C_i][C_s] \qquad \text{Equation 9}$$

Since

$$[C_f] = [C_i]$$

then

$$\frac{-dC_s}{dt} = [C_f][C_s]\,\Sigma\, k$$

The rate of hybridisation to sequence i equals $[C_f][C_s]k_i$.
Therefore,

$$\frac{\text{rate of hybridisation to sequence i}}{\text{rate of hybridisation to all sequences}} = \frac{[C_f][C_s]k_i}{[C_f][C_s]\Sigma k} \qquad \text{Equation 10}$$

$$= \frac{k_i}{\Sigma k}$$

In a hybridisation reaction, it is more likely that there will be a number (m) of cross-reacting species. The overall reaction can be treated as the sum of a number of independent hybridisations each with a different rate constant and each following pseudo-first order kinetics. When they go to completion, the probe will all be in hybrids. The fraction of the probe hybridised to sequence i is given by:

$$\frac{k_i}{\overset{m}{\underset{1}{\Sigma}} k}$$

At any time during the reaction, the ratio of the amounts of probe hybridised to sequence i and to any other filter-bound sequence j is given by the ratio of the rate constants (k_i/k_j) and the ratio is not affected by the time of incubation. Hence the discrimination between related hybrids is not affected by the extent of the reaction because all the filter-bound sequences continuously compete for the same limiting probe (*Figure 2*).

5.2.2 Single-stranded Probe in Excess

The kinetics of hybridisation are different from that described above when the probe is in excess over the filter-bound sequences, as in typical genomic Southern blots, genomic dot blots, RNA dot blots and Northern blots (Chapters 5 and 6). If the probe is single-stranded and so cannot reassociate (e.g., for M13 or SP6 RNA probes), Equation 8 simplifies to the same form as Equation 9:

$$\frac{-dC_s}{dt} = k_1[C_f][C_s] + k_i[C_i][C_s]$$

If $[C_f] = [C_i]$, then rearranging terms,

$$\frac{-dC_s}{dt} = [C_s] \overset{i}{\underset{i=1}{\Sigma}} k_i[C_i]$$

It can be shown that E_i (t), the fraction of filter-bound sequence i actually hybridised at time t, is given by the equation:

$$E_i\ (t) = 1 - e^{-k_i[C_s]t} \qquad\qquad \text{Equation 11}$$

and the ratio of the extent of hybridisation of cross-hybridising sequence i, to perfectly-matched sequence f is:

$$\frac{E_i\ (t)}{E_f\ (t)} = \frac{1 - e^{-k_i[C_s]t}}{1 - e^{-k_1[C_s]t}} \qquad\qquad \text{Equation 12}$$

This ratio is not constant but varies with time. The discrimination (that is, the actual extent of cross-hybridisation compared with the hybridisation to perfectly-matched sequences) is maximal very early in the reaction when it equals k_i/k_1, but declines with increasing incubation time as the term at the right hand side of Equation 9 approaches unity (*Figure 2*). This means that although the homologous reaction is faster and will

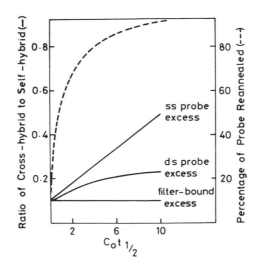

Figure 2. Effective discrimination between perfectly-matched and mismatched sequences as a function of the extent of the reaction. The solid lines represent the ratio [$E_i(t)/E_f(t)$, see text] of the amounts of probe hybridised to filter-bound mismatched, heterologous sequences (cross-hybridisation) and perfectly-matched, homologous sequences (self-hybridisation). Three separate reactions are shown: filter-bound sequences in excess; denatured double-stranded DNA probe in excess; single-stranded probe in excess. The discrimination ratio (k_i/k) is assumed to be the same (0.1) in all cases. The dashed line shows the normal kinetics of reassociation of denatured double-stranded DNA in solution. Reproduced from reference 4 with permission.

reach completion earlier, the heterologous reaction will eventually catch up (single-stranded probe excess; *Figure 2*). In practical terms, then, with increasing time of reaction, discrimination becomes poorer, so reaction times should be kept short.

5.2.3 Double-stranded Probe in Excess
Consider Equation 8 again:

$$\frac{-dC_s}{dt} = k_1[C_f][C_s] + k_2[C_s]^2 + k_i[C_i][C_s]$$

Equation 8

If the probe can reassociate, the term $k_2[C_s]^2$ is significant. Therefore, as the reaction proceeds, reassociation in solution reduces the amount of probe that is available to hybridise to the filter-bound sequences. So the reaction changes from being in probe excess to one in which filter-bound nucleic acid is in excess. It can be shown that:

$$\frac{\text{fraction of probe hybridising to sequence i}}{\text{fraction of probe hybridising to perfectly-matched sequence f}} = \frac{E_i(t)}{E_f(t)}$$

$$= \frac{1 - (1+k_2[C_s]t)^{-n}}{1 - (1+k_2[C_s]t)^{-1}}$$

Equation 13

where n = k_i/k_1, that is, the discrimination ratio. Again this means that in practical terms, discrimination equals k_i/k_1 very early in the reaction, but deteriorates rapidly. So, in practice, to distinguish between cross-hybridising species, it is best to use short times of incubation regardless of whether the probe or the filter-bound sequence is in excess. If this does not generate enough signal to be detected, it is advisable to use excess filter-bound sequence and to hybridise for longer times.

6. BINDING OF NUCLEIC ACID TO FILTERS

6.1 Types of Filter Material

There are several types of filter currently in use for the immobilisation of DNA and RNA, for example, nitrocellulose, nylon and chemically-activated papers. The material of choice depends on the purpose of the experiment.

Nitrocellulose filters bind DNA and RNA very efficiently, (~ 80 $\mu g/cm^2$) except for small fragments of less than about 500 nucleotides in length which are bound rather poorly. The binding procedure is simple. The main disadvantage of nitrocellulose is that it is rather fragile so it requires careful handling and on repeated use tends to become brittle and fall apart. Nylon filters are more pliable than nitrocellulose, are easier to handle and can be used indefinitely without disintegrating. They are reputed to bind nucleic acid as efficiently as nitrocellulose and on the whole we have found that this is true, but we have experienced batch variation in the binding efficiency of some brands. For Southern, Northern and dot blots, both nitrocellulose and nylon filters give excellent results. For DNA dot blots, filters with a pore size of 0.45 μm are used for large nucleic acid molecules and 0.22 μm for molecules of less than 500 nucleotides. For RNA dot blots, filters with $0.1-0.22$ μm pore size are most efficient.

Although nitrocellulose and nylon filters immobilise nucleic acid, binding by conventional procedures is not covalent. This can lead to problems. For example, nucleic acid is gradually leached off the surface when filters are hybridised for long periods, particularly at high temperature. Furthermore, if the probe in solution is complementary to the entire length of the filter-bound sequence, the hybrid dissociates from the filter and is lost into solution (27). So a consequence of non-covalent binding of nucleic acid is that the hybridisation sensitivity may be reduced with time. New techniques have now been developed for covalent binding of nucleic acid to membranes. This involves u.v. light-induced binding (28). However, to date this is only applicable to nylon filters because of the risk of fire when using nitrocellulose membranes. Chemically-activated paper binds nucleic acid covalently so it has the advantage that it does not discriminate against small nucleic acid molecules and does not lose nucleic acid sequences once they are bound. Both cellulose and nylon filters can be chemically activated. However, the binding capacity of chemically-activated paper ($1-2$ $\mu g/cm^2$) is much lower than other types of filter and the binding procedure is more complicated. Hence it is not much used for Southern, Northern and dot blots; its main use is in hybrid selection for the enrichment of specific RNA sequences (Chapter 5, Section 3.2).

Suppliers of nitrocellulose filters are Schleicher and Schüll (membrane filters BA85), Millipore U.K. and Waters Associates (Millipore filters), Sartorius Instruments Ltd. (Sartorius filters) and Amersham International plc (Hybond C filters). Suppliers of nylon filters are New England Nuclear (GeneScreen and GeneScreen Plus hybridisation transfer membrane), PALL (Biodyne transfer membrane) and Amersham International plc (Hybond-N membranes). Clearly, this list can never be complete since new products are continually being marketed. The filters are generally available as circles and rectangles in several sizes. Most are also available in rolls which can be cut to size.

All filters require the nucleic acid to be denatured for binding. However, it is most important to note that there is no immobilisation procedure uniformly applicable to all types of filter. Nitrocellulose filters require high ionic strength for quantitative binding of both DNA and RNA and the binding efficiency is much reduced at low ionic strength (29,30). In contrast, GeneScreen nylon membranes require low ionic strength for binding and the binding is poor at high salt concentrations. We have successfully used the same procedures for binding to nitrocellulose and to Biodyne nylon membranes.

6.2 DNA Dot Blots

Multiple samples of genomic or plasmid DNA are spotted next to each other on a single filter in dots of uniform diameter. For quantitative analysis, known amounts of DNA are applied. To evaluate the extent of hybridisation of the probe, a standard consisting of a dilution series of DNA dots is applied in an identical way to the same filter. The procedure binds samples quickly so that many samples can be handled at once. As little as $1-3$ pg of a hybridising DNA sequence can be detected. Dot blots do not distinguish the number and size of the molecules hybridising, so the hybridisation 'signal' is the sum of all sequences hybridising to the probe under the conditions used.

Commercial apparatus has been developed for binding multiple samples of DNA to filters. A protocol for use of this is described elsewhere in this volume (Chapter 5, Section 3.1). Not every laboratory has access to such a device so the procedure described here involves manual application of samples. This is more time consuming and the dots are less uniform than when applied by the multiple filtration device, but the results are perfectly satisfactory.

There are many protocols in use for binding samples to filters. They can be divided into two classes according to whether the DNA is denatured before or after it is applied to the filter. Both give satisfactory results. An example of the former method is described for use with the multiple filtration device (Chapter 5, Section 3.1.2). We shall describe an example of the latter method which is in current use in our laboratory with nitrocellulose and Biodyne filters. It is not applicable to GeneScreen filters which require low salt concentration for DNA binding.

6.2.1 *Plasmid DNA Dot Blots*

It is necessary to convert supercoiled DNA to open circular or linear form to bind to filters. This is because DNA must be single-stranded for binding and denatured supercoiled DNA renatures too quickly on neutralisation to be trapped in the denatured state. Two common ways of obtaining linear or nicked plasmid DNA are to restrict the DNA by enzymic digestion and to treat the plasmid at high temperature (see *Table 1*). The latter partially depurinates the DNA so that on subsequent treatment with alkali the phosphodiester bond breaks at the site of depurination (31). Linear DNA will then separate into single strands.

Nitrocellulose filters are usually treated with high concentrations of salt either at the

Table 1. Linearisation of Plasmid DNA.

Restriction Method

1. Digest the recombinant plasmid with a suitable restriction enzyme. Monitor linearisation of the plasmid by agarose gel electrophoresis.
2. Extract the restricted DNA with an equal volume of phenol pre-saturated with 10 mM Tris-HCl, pH 7.5, 0.1 mM NaCl, 1 mM EDTA. Spin for 2 min in a microcentrifuge to separate the layers.
3. To the aqueous phase, add 0.5 vol. of 7.5 M ammonium acetate and precipitate the DNA by adding 2.5 vol. ethanol pre-cooled to $-20°C$. Mix well and place at $-20°C$ overnight or $-70°C$ for $1-2$ h.
4. Recover the DNA by centrifugation and dry briefly under vacuum.
5. Resuspend the DNA at 50 μg/ml in TE buffer, pH 8.0[a].

High Temperature Method

1. Place 20 μg DNA in a microcentrifuge tube in a final volume of 100 μl 20 mM Tris-HCl, pH 7.4, 1 mM EDTA.
2. Pierce the lid to prevent it popping open and place the tube in a boiling water bath for 10 min.
3. Chill in ice and centrifuge for 10 sec to ensure that all the sample is at the bottom of the tube.
4. Check the volume and adjust to 100 μl with water if necessary so that the DNA concentration remains at 20 μg/100 μl.

---

[a]TE buffer is 10 mM Tris-HCl, 1 mM EDTA, pH 8.0.

same time as, or prior to, binding of nucleic acid. This both improves the efficiency of binding and helps to keep the diameter of the dot small. Salts commonly used are 1 M ammonium acetate or 20 x SSC [1 x SSC is 0.15 M NaCl, 0.015 M trisodium citrate, pH 7.0]. A suitable procedure for binding plasmid DNA to nitrocellulose filters involving pre-treatment of the filter with high salt is given below. It is important to note that the filter must not at any stage be handled with bare hands. Grease from the fingers will result in poor binding of nucleic acid and high backgrounds. Therefore disposable plastic gloves must be worn at all stages.

(i)   Float a sheet of nitrocellulose on water taking care not to trap air bubbles under-neath. When one side is wet, immerse the filter completely to wet the other. If there are dry patches which are reluctant to wet, boil the water for a few minutes.

(ii)   Blot the filter lightly on Whatman 3MM paper and transfer to a dish containing 20 x SSC. Leave for 30 min with gentle shaking.

(iii)   Dry at room temperature or under a lamp until completely dry.

(iv)   If convenient, use a conventional rubber stamp and ink pad to stamp the paper with an array of 5 mm diameter circles to allow easy identification of samples. At this stage, the filters can be stored dry at room temperature sealed in a poly-thene sleeve.

(v)   Place a nitrocellulose filter which has been treated with 20 x SSC onto the lid of a plastic box such that only the edges of the filter are in contact with the lid.

(vi)   Apply the plasmid (0.8 μg in 4 μl, linearised as in *Table 1*) to the filter. This can be achieved using a $1-5$ μl Supracap pipette (Brand) or an automatic Pipet-man (Gilson). Be careful not to puncture the nitrocellulose filter. Keep the diam-eter of the dots small and do not exceed 4 mm. If necessary, make repeated applications allowing time for each application to dry.

(vii)   Allow the samples to dry at room temperature or under a lamp.

Table 2. Reducing the Size of Eukaryotic DNA Prior to Filter Binding.

1.	*Either* sonicate 30 μg DNA to an average size of 2 kb or restrict it with a suitable restriction enzyme and then remove the enzyme by phenol extraction (*Table 1*, step 2). The size of the DNA should be checked by agarose gel electrophoresis using appropriate size markers (Appendix II).
2.	Recover the DNA by ethanol precipitation (*Table 1*, step 3).
3.	Wash the DNA in 70% ethanol, dry briefly under vacuum and resuspend in 1 ml TE buffer, pH 8.0[a].
4.	Measure the concentration of DNA spectrophotometrically using the conversion factor $A_{260nm} = 1$ for a solution of 50 μg/ml.
5.	Check the volume of solution and freeze dry.
6.	Resuspend the DNA in water at a concentration of 10 μg per 4 μl. For serial dilutions, prepare a set of microcentrifuge tubes each containing 6 μl TE buffer, pH 8.0. Remove 6 μl DNA into the first tube containing TE buffer. Mix well. Remove 6 μl from this tube into the second tube with TE buffer and so on until seven dilutions have been made (10 μg−5.3 ng per 4 μl).
7.	Centrifuge briefly in a microcentrifuge to ensure that each DNA solution is at the bottom of the tube.
8.	Apply the samples to a dry sheet of nitrocellulose pre-treated with 20 x SSC as described in Section 6.2.1, steps (v) − (xii).

[a]TE buffer is 10 mM Tris-HCl, 1 mM EDTA, pH 8.0.

(viii)   Denature the DNA by placing the filter, application side up, on a sheet of What-man 3MM paper saturated, but not 'swimming', in 1.5 M NaCl, 0.5 M NaOH. Leave for 5 min. (This is conveniently done in a plastic tray.)

(ix)   Transfer the filter to Whatman 3MM paper saturated with 0.5 M Tris-HCl, pH 7.4, for 30 sec.

(x)   Transfer the filter to Whatman 3MM paper saturated in 1.5 M NaCl, 0.5 M Tris-HCl, pH 7.4, for 5 min. The DNA is now reversibly bound to the filter.

(xi)   Place the filter on a dry sheet of 3MM paper and leave to dry at room temperature.

(xii)   Sandwich the filter between two sheets of 3MM paper and bake at 80°C for 2 − 3 h to immobilise the DNA. Ideally, nitrocellulose filters should be baked in a vacuum oven to reduce the risk of fire.

6.2.2 *Genomic DNA Blots*

Like plasmid DNA, genomic DNA can be applied to filters in dots. The amount of DNA required per dot depends on the experiment being performed. To detect a single copy sequence in eukaryotic DNA, a minimum of 10 μg DNA per dot is suggested. For sequences present in multiple copies, proportionately less can be used.

Modern techniques for DNA isolation usually give a product which is very concentrated and has a very high molecular weight. These two factors make the DNA solution very viscous so that it is difficult to measure the concentration accurately. Hence, to do this, it is necessary to reduce the size of the DNA either by sonication or digestion with an appropriate restriction endonuclease (*Table 2*). This also helps to bind the DNA to the filter more efficiently. Binding of the DNA to nitrocellulose filters, pre-treated with 20 x SSC, is carried out as described above for plasmid DNA (Section 6.2.1).

6.3 **RNA Dot Blots**

The principle of this procedure is exactly the same as for DNA dot blots. Known amounts of RNA are applied to an inert support and the amount of specific RNA sequence is determined by hybridisation with a suitable labelled probe. Evaluation of the extent of hybridisation can be made by comparison with standards. The technique is sensitive — as little as 1 pg of a specific RNA sequence can be detected (32). As with DNA dot blots, however, the procedure gives no information on the size or number of sequences contributing to the hybridisation signal. Nylon and nitrocellulose filters are suitable supports. Because nitrocellulose tends to be the most used, its use will be described here. Chemically-activated paper is not generally used for RNA dot blots as its binding capacity is too low.

6.3.1 *Preliminary Precautions*

One of the main problems of working with RNA is its extreme sensitivity to degradation. Glassware must be scrupulously clean and should never be touched with bare hands which are a good source of ribonuclease. Prior to use, the glassware should be treated with diethylpyrocarbonate to inactivate any ribonuclease. This can be done by immersing the glassware in water to which has been added two drops per litre of diethylpyrocarbonate and then boiling for 30 min. The glassware can then be dried. Heavy metal ions can lead to degradation of RNA especially when long incubations are involved so these should be removed by filtration of all solutions through Chelex resin (BioRad) before use.

6.3.2 *Denaturation of RNA and Binding to Filters*

Although RNA is single-stranded, it contains double-stranded regions which must be denatured for efficient binding to filters. Alkali treatment is not suitable since it degrades RNA and heat denaturation has not been found to give efficient binding (32). Commonly used denaturants for RNA include glyoxal (33), methyl mercuric hydroxide (34), formaldehyde (35) and dimethyl sulphoxide (DMSO) (36). Since methyl mercuric hydroxide and formaldehyde are toxic and DMSO dissolves nitrocellulose, the procedure described here uses glyoxal as a denaturant. It is based on the methods developed by Thomas (32,37). Glyoxal is supplied commercially as a 40% aqueous solution (6.89 M) which contains polymerisation inhibitors. Glyoxal is readily oxidised to glyoxilic acid which degrades RNA so it is necessary to purify the glyoxal before treating the RNA. This is usually done by deionisation. A suitable protocol is as described in Chapter 6, *Table 1*.

Glyoxal denatures RNA (and DNA) by binding covalently to guanine residues forming an adduct which is stable at acid and neutral pHs. Glyoxylated nucleic acid binds efficiently to nitrocellulose paper but, after binding, the glyoxal groups must be removed because they have an inhibitory effect on hybridisation. This is easily and quantitatively achieved by treating the filter at 100°C at pH 8.0. Under these conditions as little as 1 pg of a specific sequence of RNA can be detected. The detailed procedure for glyoxalation of RNA and binding to nitrocellulose filters is as follows.

(i)   Dry down 20 μg of the sample RNA in a microcentrifuge tube and dissolve it in 5 μl water. The RNA should be salt-free and free of protein which will otherwise react with the glyoxal.

(ii)   Prepare a denaturation solution: 34 μl deionised glyoxal, 20 μl 0.1 M sodium phosphate buffer, pH 6.5, 46 μl water.

(iii) Add 5 μl denaturation solution to the RNA. Cover the tube and incubate for 1 h at 50°C. (If using a water bath, make sure that the water comes well up the sides of the tube in order to minimise evaporation.)

(iv) Centrifuge in a microcentrifuge (10 sec) to ensure the sample is at the bottom of the tube. After denaturation in glyoxal, the samples are stable for a few hours and can be kept at 4°C.

(v) If required, make serial dilutions as described in *Table 2* (step 6) but using 1% SDS as the diluent.

(vi) Apply the samples by hand to a sheet of nitrocellulose [pre-treated in 20 x SSC as described in Section 6.2.1, steps (i)–(iv)] using the application procedure described in Section 6.2.1, steps (v) and (vi).

(vii) Bake the filter at 80°C for 2 h to immobilise the RNA.

(viii) Remove the glyoxal groups by placing the filter in water at 100°C for 5–10 min, then allowing the water to cool to room temperature.

(ix) Blot the filter on 3MM paper and allow to dry at room temperature or under an infra-red lamp.

7. NUCLEIC ACID PROBES

7.1 Types of Probe

In theory, any nucleic acid can be used as a probe provided that it can be labelled with a marker which allows identification and quantitation of the hybrids formed. In practice, double- and single-stranded DNAs, mRNA and RNAs synthesised *in vitro* are all used as probes. Oligonucleotide probes are not used in quantitative dot blots; they are most useful for screening recombinant DNA libraries.

7.1.1 *Double-stranded DNA Probes*

Double-stranded DNA probes are very commonly used in dot blot analysis. They are often cloned sequences and have low complexity. There are two important points to note when using double-stranded DNA probes:

(i) Two competing reactions occur in filter hybridisation, *viz.* reassociation of the probe in solution and hybridisation to the filter-bound nucleic acid. Therefore, reaction conditions should be chosen to optimise the latter (see Section 11).

(ii) If the DNA is a cloned sequence, it should be excised and purified away from the vector. This avoids complications which can arise if single-stranded vector tails allow formation of concatenates in solution, particularly if the DNA has been randomly sheared. Furthermore, if filters are re-hybridised and the previous probe containing vector sequences has not been completely removed, sandwich hybridisation may occur. That is, duplexes may form through vector sequences rather than through insert sequences. This complicates the interpretation of results.

7.1.2 *Single-stranded DNA Probes*

With single-stranded DNA probes there is no competing reassociation in solution so filter hybridisation is favoured and reactions can be carried out for longer. Single-stranded DNA probes are obtained by strand separation of double-stranded DNA (see Chapter 6, Section 4.2.4) or from M13 phage recombinants.

7.1.3 *RNA Probes*

RNA probes are more difficult to handle than DNA probes because of the widespread presence of ribonucleases. In addition, mRNA probes, or cDNAs derived from them, are often complex mixtures of sequences and therefore the sequence of interest may represent only a very small proportion of the total nucleic acid. Since the rate of filter hybridisation is inversely proportional to the complexity for low amounts of filter-bound nucleic acid, it may be difficult or impossible to detect the desired hybrids.

Recently it has proved possible to synthesise large amounts of RNA *in vitro* from specially-constructed recombinant plasmids, such as the SP6 plasmids. The probes have low complexity and, because they are single-stranded, there is no competing reassociation reaction in solution. For these reasons, the use of SP6 RNA transcripts as probes is proving increasingly popular.

7.2 **Radiolabelled Probes**

Traditionally, filter hybridisations have been carried out with radioactively-labelled probes. ^{32}P is the most commonly used radionuclide and will be the only one discussed here. Conventional labelling replaces a proportion of the nucleotides in a nucleic acid with ^{32}P derivatives or adds ^{32}P to the end of the molecule. After hybridisation, hybrids are detected by autoradiography. ^{32}P has the advantage over other radioisotopes that high specific activities can be readily attained. Much of the technology of filter hybridisation has been developed with it. However, precautions must be taken when handling ^{32}P because of the radiation emitted. Detection by autoradiography, while sensitive, may take a long time if there are few counts in the hybrids. Furthermore, since ^{32}P has a half-life of 14.3 days, experiments should be completed within one half-life.

The preparation of radioactively-labelled DNA and RNA probes, including SP6 RNA transcripts, is described in detail in Chapter 2. In preparing labelled probes for filter hybridisations it is important to remove unincorporated precursors efficiently before use (see Chapter 2) otherwise they may bind non-specifically, but irreversibly, to the filter, giving a high background.

7.3 **Non-radioactive Probes**

Recent advances in nucleic acid technology now offer alternatives to radioactively-labelled probes. For example, single-stranded DNA can be coupled to a protein. If this protein-DNA complex is now hybridised to filter-bound nucleic acid, the protein in the duplex can be visualised by an antibody reaction (38). If the protein is an enzyme such as peroxidase, then it can be detected and quantitated by its ability to convert a colourless substrate into an insoluble coloured pigment at the site of hybrid formation. This technique is sensitive (1 – 5 pg nucleic acid can be detected) and has some potentially useful applications. For example, DNA probes coupled to different enzymes can be used in the same hybridisation reaction, so that it should be possible to detect the presence of unrelated sequences simultaneously (38).

Another procedure that uses non-radioactive probes and is becoming increasingly popular is biotin labelling of nucleic acid (39,40). These probes are prepared in a nick-translation reaction by replacing nucleotides with biotinylated derivatives. After hybridisation and washing, detection of hybrids is by a series of cytochemical reactions which finally give a blue colour whose intensity is proportional to the amount of biotin in the hybrid. Biotinylated probes detect target sequences with the same sensitivity as radioactive probes, that is, in the $1-5$ pg range. There are several advantages of using biotinylated probes. For example, non-toxic materials are employed and there are no problems of inconveniently short half-lives of the label. This has the additional bonus that biotin-labelled probes can be prepared in advance in bulk and stored at $-20°C$ until required. Detection of hybrids is much faster than for radioactive probes, visualisation of hybrids being complete $2-4$ h after washing. One disadvantage of biotin-labelled probes is that the cytochemical visualisation reactions lead to precipitation of insoluble material which cannot be removed, so when the filter is re-used, the previous 'signals' are still present (39,40). The preparation and use of non-radioactive nucleic acid probes is discussed in Chapter 2, Section 4.3.

7.4 Additional Considerations

Additional factors which should be borne in mind when choosing probes are:

(i)   It is important to characterise the nucleic acid used for the probe. If any repetitive elements are present (e.g., *Alu*I sequences), they must be removed if the probe is to be used to detect low copy number sequences otherwise hybridisation of the latter will be masked by the repetitive sequence hybridisation.

(ii)   The length of the labelled probe is important since the kinetics of hybridisation depend on probe length (see Section 3.3).

(iii)   The kinetics of hybridisation differ according to whether the probe or filter-bound sequences are in excess and it is not always immediately apparent which is in excess. What is important is the concentration of the hybridising species, not the total nucleic acid concentration. The following is a rough guideline to this problem. With genomic Southern and Northern blots and genomic DNA and RNA dot blots, the concentration of the probe is likely to be in excess. For example, in *Figure 3*, even though there is 10 μg RNA per dot and the Ha-*ras*1 probe is at 20 ng/ml, the probe is in excess for RNA taken from normal tissue. (However, note that the filter-bound sequences are in excess for RNA taken from diseased tissue.) With plasmid or phage dot blots, and phage and colony screening, the filter-bound sequences are likely to be in excess. However, to check which is in excess, the following preliminary experiments can be performed.

(a) Vary the input of probe; if the filter-bound sequence is in excess, the amount of hybridisation should be proportional to the probe input.

(b) Vary the amount of nucleic acid on the filter; if it is in excess, there should be no difference in the amount of probe hybridised.

8. HYBRIDISATION USING RADIOACTIVE PROBES

8.1 Choice of Reaction Conditions

There are many protocols available for hybridising a probe in solution to nucleic acid immobilised on filters. The conditions used depend on the purpose of the experiment and in general are governed by whether DNA-DNA or DNA-RNA hybridisation is involved and whether closely-related or distantly-related sequences are reacting. Reaction conditions that permit formation of hybrids which have a high degree of mismat-

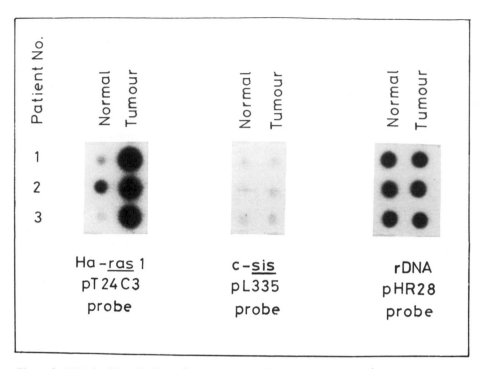

Figure 3. RNA dot blots. Replicate filters containing 10 μg per dot of poly(A)$^+$ RNA from normal and tumour breast tissue of three different patients were hybridised with the following denatured ^{32}P-labelled double-stranded probes: pT24C3 (Ha-*ras*-1), pL335 (c-*sis*) and pHR28 (rDNA). Equal amounts of RNA were present in the dots as judged by the intensities of the hybridisation signals using the rDNA probe. Transcripts homologous to the Ha-*ras*-1 probe are more abundant in tumour tissue compared with normal breast tissue and this difference is specific to the *ras* oncogene since the c-*sis* probe did not give a signal. (Data of Agnatis and Spandidos, with permission.)

ching are said to be permissive (relaxed or low stringency), while those which allow only well-matched hybrids to form are said to be stringent (high stringency). However, the same basic procedure is followed irrespective of the particular reaction conditions used. We will first describe standard hybridisation protocols which have widespread applicability (Sections 8 – 10) and then how varying the reaction conditions can be exploited to detect different hybrids (Section 11).

The hybridisation process can be divided into three steps: pre-hybridisation, hybridisation with a labelled single-stranded probe, and washing. In the *pre-hybridisation* step, the filter is incubated in a solution which is designed to pre-coat all the sites on it which would bind the probe non-specifically. Failure to do this leads to high backgrounds. Typically, the solution contains Ficoll, polyvinyl pyrollidone and bovine serum albumin [i.e. Denhardt's solution (ref. 41)], and heterologous DNA. As an alternative, heparin can substitute for Denhardt's solution (42). To reduce backgrounds even further, poly(A) and poly(C) are often included. Poly(A) is useful when the probe or filter-bound sequences are rich in A and T residues, e.g., poly(A)$^+$ mRNA or cDNA derived from it. Similarly, poly(C) is included if the probe or filter-bound sequences are rich in G and C residues as when a recombinant is generated through oligo(dG) and oligo(dC) homopolymer tailing. For hybridisations involving RNA, yeast tRNA is often used as a competitor. For *hybridisation*, it is necessary to ensure that the added nucleic acid probe is single-stranded. For double-stranded DNA probes this is usually achieved by boiling or by denaturing in alkali. Radioactive probes can be denatured using either method; heat denaturation is described here and alkaline denaturation is described in Chapter 5 (*Table 3*, step 6). For most purposes hybridisation can be carried out in either aqueous solution or in the presence of formamide. We use the same formamide-containing solutions for both RNA-DNA and DNA-DNA hybridisations, but aqueous solutions for DNA-DNA hybridisations only. Both protocols can be used with nitrocellulose and nylon filters and are described below. After hybridisation, *washing* is carried out to remove unhybridised probe and to dissociate unstable hybrids. The temperature and salt concentration of the washing solution determine which hybrids will be dissociated. In general, washing should be under as stringent conditions as possible; at $5-20°C$ below T_m for well-matched hybrids and $12-20°C$ below T_m for cross-hybridising species. In practice, $65-70°C$ is usually chosen for hybrids having a high degree of homology and $50-60°C$ for poorly-matched hybrids.

Where possible, the pre-hybridisation, hybridisation and washing steps should be carried out in a shaking water bath or on a shaking platform in an incubator. In filter-bound nucleic acid excess, diffusion of the probe to the filter can be limiting in the absence of agitation. Also, high backgrounds are sometimes encountered if there is no shaking. Solutions should be pre-warmed to the required temperature prior to use.

8.2 Hybridisation in the Presence of Formamide

8.2.1 *Pre-hybridisation*

(i) To wet the filters evenly, float them on a solution of 1% Triton X-100, taking care to prevent air bubbles being trapped underneath. When one side is wet, immerse the filter to wet the other side.

(ii) Remove the filters and blot gently on Whatman 3MM paper to remove excess liquid.

(iii) It is convenient to carry out the pre-hybridisation and hybridisation reactions in the same container. Typically this is a polythene bag. Suitable bags are 'Sears Seal-N-Save Boilable Cooking Pouches' or Layflat polythene tubing (Trans-Atlantic Supplies). Place each wet filter in a separate bag and heat-seal this, except for one corner, using a domestic bag sealer.

(iv) Add the pre-hybridisation solution (0.08 ml/cm² of filter) which has been prepared as described in *Table 3* and pre-warmed to 42°C. Gently squeeze out the air bubbles and heat seal the corner.

(v) Incubate the filter in the bag for $4-24$ h at 42°C. This can be done by placing the bag in a box of water at 42°C in a shaking water bath at the same temperature. Set the water bath to shake at a speed such that the liquid in the bag sweeps gently over the surface of the filter. Alternatively, place the bag on a shaking platform in an incubator at 42°C.

(vi) Cut a corner of the bag and drain the liquid out. Roll a pipette over the surface of the bag to remove as much of the liquid as possible. However, it is most

Table 3. Preparation of Pre-hybridisation Buffer Containing Formamide.

Solution A

Mix together:

Deionised formamide[a]	50 ml
20 x SSC[b]	25 ml
100 x Denhardt's solution[c]	5 ml
1 M sodium phosphate buffer, pH 6.8[d]	5 ml
20% SDS[e]	0.5 ml

Adjust the volume to 95.5 ml with water.

Solution B

Mix together:

Sonicated calf thymus DNA or salmon sperm DNA at 5 mg/ml[f]	2 ml
Poly(C) [5 mg/ml][g]	0.2 ml
Poly(A) [5 mg/ml][g]	0.2 ml
Yeast tRNA (5 mg/ml)[g]	2 ml

Denature in a boiling water bath for 5 min.
Quench in ice.

Add solution B to solution A and store at 4°C.

---

[a]Formamide is a teratogen. Handle with care and use gloves. All contaminated glassware should be soaked overnight in dilute H_2SO_4 then rinsed with water before washing as usual. To deionise formamide, add 200 ml formamide to ~ 10 g of AG501-X8(D) mixed-bed resin (Bio-Rad). Stir for 1 h at room temperature. Filter through Whatman No. 1 filter paper to remove the resin. Store at 4°C in a dark bottle.
[b]The composition of SSC is 0.15 M NaCl, 0.015 M trisodium citrate, pH 7.0.
[c]100 x Denhardt's solution contains 2% Ficoll (mol. wt. 400 000), 2% polyvinyl pyrrolidone (mol. wt. 400 000) and 2% bovine serum albumin. Store at −20°C.
[d]1 M sodium phosphate buffer, pH 6.8, is made by mixing 25.5 ml of 1 M NaH_2PO_4 and 24.5 ml of 1 M Na_2HPO_4. Store at room temperature.
[e]Store this stock solution at room temperature.
[f]Add the DNA to water at ~ 5 mg/ml. Stir; it may take several hours for the DNA to dissolve. Then sonicate to a length of 400 − 800 bp. The size can be checked by agarose gel electrophoresis. Adjust the concentration to 5 mg/ml [A_{260nm} = 1 for a solution of 50 μg/ml]. Store at −20°C.
[g]These solutions are stored at −20°C. Their addition to the pre-hybridisation buffer is optional (see Section 8.1). However, there is no disadvantage in adding them even if T- and G-rich sequences are not present in the filter-bound nucleic acid sequence.

important that the filter is not allowed to dry out if high backgrounds are to be avoided. Therefore the filters should be left in the pre-hybridisation buffer in the bag until just before applying the hybridisation solution.

8.2.2 *Hybridisation*

(i)   Except for single-stranded probes such as RNA and M13 probes, denature the labelled probe by placing it in a boiling water bath for 5 min. Quench in ice.

(ii) The hybridisation can be carried out in the presence of dextran sulphate which increases the rate of hybridisation (Section 3.8) or in its absence. For hybridisation buffer containing dextran sulphate (prepared as described in *Table 4*), pre-warm the buffer and add the denatured probe to a concentration which does not exceed 10 ng probe/ml or high backgrounds may ensue. In the absence of dextran sulphate, the probe concentration can be increased to 50 − 100 ng (43). For radioactive probes which have been labelled to a specific activity of

Table 4. Preparation of Standard Hybridisation Solution Containing Formamide.

Solution A

Mix together:

Deionised formamide[a]	50 ml
20 x SSC[b]	25 ml
100 x Denhardt's solution[c]	1 ml
1 M sodium phosphate buffer, pH 6.8[d]	2 ml
20% SDS[e]	1 ml
Dextran sulphate (mol. wt. 500 000)[h]	10 g

Stir until the dextran sulphate has dissolved. Adjust the volume to 95.5 ml.

Solution B

Mix together.

Sonicated DNA (5 mg/ml)[f]	2 ml
Poly(C) [5 mg/ml][g]	0.2 ml
Poly(A) [5 mg/ml][g]	0.2 ml
Yeast tRNA (5 mg/ml)[g]	2 ml

Denature in a boiling water bath for 5 min.

Quench in ice.

Add solution B to solution A and store at 4°C.

[a-g]See corresponding footnotes to *Table 3*.
[h]The inclusion of dextran sulphate is optional (see text, Section 3.8).

$1-2$ x 10^8 c.p.m./μg, a probe concentration of 10 ng/ml gives a solution of about $1-2$ x 10^6 c.p.m./ml).

(iii) Immediately add this solution to the filter (0.05 ml/cm² filter) and reseal the bag.

(iv) Hybridise the filter at 42°C for the required time. This is normally between 6 and 48 h. Overnight is convenient and for many purposes is sufficient, but see Section 11.2.

8.2.3 *Washing*

(i) Cut one corner from the bag and remove the hybridisation solution. Retain the probe if it is to be re-used (see Section 13), otherwise discard it down a designated sink.

(ii) Cut open the bag completely and immerse the filter in 200 ml of 2 x SSC, 0.1% SDS at room temperature. Shake the filter gently. Rinse the filter twice for 5 min each time in this solution.

(iii) For a moderately stringent wash, wash the filter twice in 400 ml of 2 x SSC, 0.1% SDS at 60°C for 1 h. For a higher stringency wash, treat the filter for 2 x 1 h at 65°C in 0.1 x SSC, 0.1% SDS.

(iv) Finally, rinse the filter in 2 x SSC at room temperature. Blot the filter to remove excess liquid but do *not* dry the filter if it is to be re-washed or re-screened (Section 13).

(v) Detect the hybrids as described in Section 10.1.

8.2.4 *Washing with Nuclease Treatment*

In principle, blots can be treated with nucleases to remove unpaired loops and single-stranded probe tails. This practice should increase the specificity of the reaction, but it has not been studied systematically and there is no real evidence that the treatment is effective. It is better to control specificity by careful choice of reaction and dissociation conditions than through enzyme digestion. DNase and nuclease S1 treatments are not generally used in filter hybridisations, but RNase treatment occasionally is. Filters are treated with a mixture of RNase A and T1 RNase at 25 μg/ml and 10 units/ml, respectively, in 2 x SSC at 37°C for 2 h, then washed in 2 x SSC, 0.5% SDS at 68°C, and finally in 2 x SSC at room temperature.

8.3 Hybridisation in Aqueous Solution

8.3.1 *Pre-hybridisation*

(i)   Wet the filter in 1% Triton X-100 and blot to remove excess liquid.
(ii)   Immerse in 4 x SET buffer for 15 min at room temperature. The composition of 4 x SET buffer is 0.6 M NaCl, 1 mM EDTA, 80 mM Tris-HCl, pH 7.8.
(iii)   Transfer the wet filter to a plastic bag and heat seal this except for one corner (see Section 8.2.1). Add pre-hybridisation buffer (0.08 ml/cm^2 of filter), prepared as described in *Table 5* and pre-warmed to 68°C. Incubate at 68°C for between 2 and 16 h.
(iv)   Open the bag and remove the pre-hybridisation buffer (see Section 8.2.1 for methodology).

8.3.2 *Hybridisation*

(i)   Denature the probe as described in Section 8.2.2 and add it to fresh pre-hybridisation buffer (pre-warmed to 68°C) at $10-25$ ng/ml. Add this solution to the filter at 0.05 ml/ cm^2 of filter.
(ii)   Incubate at 68°C for between 5 and 16 h.

Table 5. Preparation of Aqueous Pre-hybridisation Buffer.

Solution 1	
Mix together:	
20 x SET buffer[a]	20 ml
100 x Denhardt's solution[b]	10 ml
20% SDS[b]	0.5 ml
5% sodium pyrophosphate	0.1 ml
Adjust the volume to 97.5 ml with water.	
Solution 2	
Mix together:	
Sonicated DNA (5 mg/ml)[b]	2 ml
Poly(C) [5 mg/ml][b]	0.2 ml
Poly(A) [5 mg/ml][b]	0.2 ml
Denature in a boiling water bath for 5 min.	
Quench in ice.	
Add to solution 2 to solution 1 and store at 4°C.	

[a]1 x SET = 0.15 M NaCl, 20 mM Tris-HCl, pH 7.8, 1 mM EDTA.
[b]See relevant footnote to *Table 3*.

8.3.3 *Washing*

(i)   Cut one corner from the bag and remove the hybridisation solution. Retain the probe if it is to be re-used (see Section 13), otherwise discard it down a designated sink.

(ii)   Cut open the bag completely. Remove the filter and immerse it in 4 x SET buffer, 0.1% SDS, 0.1% sodium pyrophosphate for 5 min at room temperature.

(iii) Wash the filter three times in 2 x SET, 0.1% SDS, 0.1% sodium pyrophosphate for 20 min each wash at 68°C.

(iv)   For a moderately stringent wash, wash the filter three times in 1 x SET, 0.1% SDS, 0.1% sodium pyrophosphate for 20 min each wash at 68°C. For a higher stringency wash, replace the 2 x SET by 0.1 x SET.

(v)   Rinse the filter for 5 min in 2 x SET, at room temperature.

(vi)   Blot the filter to remove excess liquid but do *not* dry the filter if it is to be re-washed or re-screened (Section 13).

(vii) Detect the hybrids as described in Section 10.1.

9. HYBRIDISATION USING BIOTIN-LABELLED PROBES

The hybridisation procedure for biotin-labelled probes is essentially the same as for radioactively-labelled probes (Section 8) except that the following points should be noted (39,40):

(i)   The probe should be denatured at high temperature and not with alkali because the amide bond in the linker molecule between the biotin and nucleic acid is alkali-labile.

(ii)   Hybridisation is carried out in solutions containing formamide rather than at high temperature. The thermal stability of biotin-labelled hybrids is slightly lower than that of radioactive hybrids. So, in practice, the formamide concentration is lowered from 50% to 45% in otherwise standard hybridisation conditions.

(iii) Certain types of polythene bags are not suitable for hybridisation with biotin-labelled probes as they lead to high backgrounds. Layflat polythene tubing (Trans Atlantic Supplies) and Sears' Boilable Cooking Pouches are both suitable.

(iv)   Since very low background signals are obtained with biotinylated probes, the concentration of probe can be increased to 250 − 750 ng/ml in the hybridisation solution. This has the additional advantage of allowing short hybridisation times of 1 − 2 h.

10. DETECTION AND QUANTITATION OF HYBRIDS

The detection of hybrids involving probes labelled with non-radioactive markers is described in Chapter 2, Section 4.3. Here we shall consider only radioactive hybrids.

10.1 **Detection**

For detecting [^{32}P]hybrids, autoradiography is the most commonly used technique. It is sensitive, gives good resolution and does not involve destruction of the filter.

If the filter is not to be re-screened or re-used, dry it at room temperature or under an infra-red lamp. Then expose it to X-ray film (e.g., Kodak X-Omat RP) at room temperature in a light-proof cassette. The time of exposure will vary from several hours to 14 days depending on the level of radioactivity in the hybrids. As a rough guide, a dot containing 100 c.p.m. ^{32}P will give a good signal on X-ray film with an overnight exposure. If the radioactivity levels are low, use of an intensifying screen (e.g., Ilford fast tungstate or Fuji Mach II) increases the sensitivity of the film by 4- to 5-fold. The film is sandwiched between the filter and the intensifying screen in the cassette. Exposure is at $-70°C$ because fluorescence reflected off the intensifying screen is prolonged at low temperatures. If two intensifying screens are used, the sensitivity of the film is enhanced $8-10$ times. In this case the filter and film are sandwiched between the two intensifying screens. For low levels of radioactivity, the film can be pre-flashed and placed flashed side against the intensifying screen with the filter on top (Chapter 5, *Table 3*, step 15). As few as $5-10$ c.p.m. per dot above background can be evaluated dependably by this adaptation of autoradiography.

If the filter is to be re-used (Section 13.1), it must not be allowed to dry otherwise the probe will bind irreversibly. The wet filter is covered in Clingfilm or Saranwrap or is inserted into a thin polythene sleeve before exposure to X-ray film. The filter should not be too wet or ice crystals will form when the cassette is placed at $-70°C$. This will distort the filter and could cause it to crack.

10.2 Quantitation of Hybridisation Signals

For many purposes it is sufficient to compare visually the intensity of hybridisation signal on an autoradiogram with that generated by a standard series of dots. The accuracy is better than 2-fold over a 100-fold range, taking into account both the intensity and diameter of the autoradiographic spots. For example, *Figure 4* shows dot blots of genomic DNA from patients suffering from chronic myeloid leukaemia (CML) and from two CML cell lines, K562 and NALM-1, probed with the c-*sis* and c-*abl* oncogenes and a human immunoglobulin λ light chain variable gene sequence (IgV$_\lambda$). Visual comparison of the autoradiographic signals indicates that cell line K562 contains about four times more copies of the c-*abl* and IgV$_\lambda$ genes than the other cell lines.

For more accurate quantitation, densitometry can be used. This is a very simple and sensitive procedure; as little as $5-10$ c.p.m. above background can be evaluated reliably. It is the best method of quantitation when the amount of radioactivity in hybrids is low. A scan is made of a series of standard dots and of the unknown samples. The area under the peaks is integrated, either electronically or the peaks can be cut from paper traces and weighed. The weight of the paper is a measure of the autoradiographic signal. A graph is then plotted of the weight of (or area under) the standard peaks against the known amount of nucleic acid on the filter. The concentration of the probe must be in excess over that on the filter and the autoradiograph should not be overexposed. An example of densitometric quantitation of a blot is given in *Figure 5*. Note that the curve relating intensity of signal (area under the peak) to the amount of RNA in the dot is only linear for a restricted range of amounts of filter-bound RNA. So, for the probe used in *Figure 5*, quantitation can be carried out only over the range $0-6$ μg RNA per dot since beyond this the filter-bound sequences are in excess of this probe. The curve in *Figure 5c* does not reach a plateau in the range analysed because the size of the dots is not uniform.

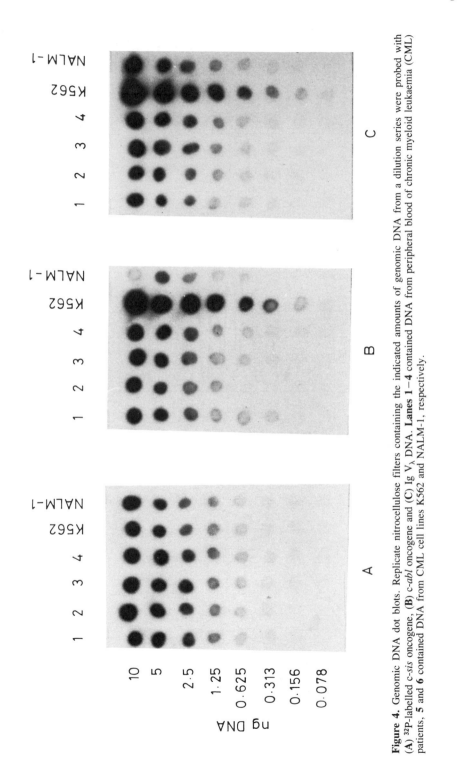

Figure 4. Genomic DNA dot blots. Replicate nitrocellulose filters containing the indicated amounts of genomic DNA from a dilution series were probed with (**A**) ^{32}P-labelled *c-sis* oncogene, (**B**) *c-abl* oncogene and (**C**) Ig V_λ DNA. **Lanes 1–4** contained DNA from peripheral blood of chronic myeloid leukaemia (CML) patients, **5** and **6** contained DNA from CML cell lines K562 and NALM-1, respectively.

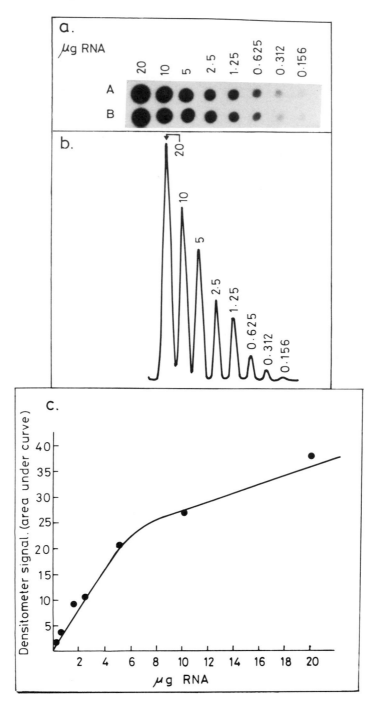

Figure 5. Quantitation of RNA dot blots. **(a)** Serial dilutions of RNA from patient 2 in *Figure 3* were prepared and applied in duplicate (A,B) to a nitrocellulose filter. The filter was hybridised with the Ha-*ras*-1 probe (*Figure 3*). **(b)** Densitometric scan made across lane A. **(c)** The relationship between the amount of RNA in each dot and the area under the densitometric peak for each dot.

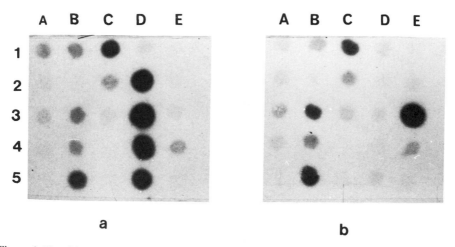

Figure 6. Plasmid DNA dot blots. Duplicate filters contained cloned recombinant plasmid cDNAs derived from mRNA of a patient suffering from acute non-lymphocytic leukaemia (ANLL). Probes were [^{31}P]cDNA derived from unfractionated mRNA of (**a**) ANLL and (**b**) chronic lymphocytic leukaemia (CLL) patients. The DNA bound to the filter was in significant excess over the probe. Data from M. Warnock, with permission.

If the hybrids are sufficiently radioactive, the dots can be cut out and counted in a liquid scintillation counter. This means, of course, that the filter cannot be re-hybridised.

10.3 Quantitative Analysis of Nucleic Acid Complexity

As we have seen, the rate of hybridisation is inversely proportional to the complexity of nucleic acid for both solution hybridisation and nucleation-limited filter hybridisation although not for diffusion-limited filter hybridisation (Section 2). Reassociation kinetics in solution have been used extensively to analyse the complexity of DNA and RNA populations and it might be supposed that nucleation-limited filter hybridisation could be used for a similar purpose. However, filter hybridisation is not suitable for quantitative studies of complexity. This is because the rate of filter hybridisation is so low that it is difficult to obtain C_0t values high enough for single copy sequences to hybridise (see Section 2).

10.4 Measurement of Relative Abundance of RNA Transcripts

For high and medium-abundance classes, dot blot hybridisation can be used to measure the relative prevalences of different mRNA species (44,45). Cloned recombinant cDNAs are applied in dots to filters and hybridised with either labelled mRNA or the cDNA derived from it. Filter-bound DNA is in excess so the extent of hybridisation is a measure of the concentration of the cloned cDNA sequence in the mRNA probe. [The extent of reaction is a reproducible characteristic of each clone and not a function of the cloned insert length, at least between the limits of 400 − 1500 nucleotides tested (44,45)]. It is estimated that a clone must be represented to a level of at least 0.1% of the mass of mRNA to be detected (44,46). This is probably true for optimal reaction conditions, but in practice the lower limit is more likely to be nearer 0.5%. An example is shown in *Figure 6*. Recombinant cDNA clones were constructed using mRNA from a patient

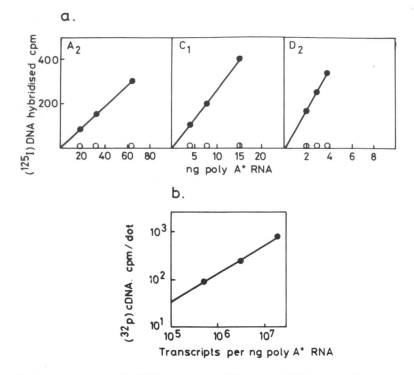

Figure 7. Prevalence analysis of mRNA transcripts. (a) Titration of cDNA clones with poly (A)+ mRNA. Complementary strands (open and closed circles) of [125I] cDNA recombinant plasmids A2, C1, and D2 (*Figure 6*) are hybridised separately in solution with the indicated amounts of poly(A)+ RNA from ANLL cells. Reactions are carried out to 20 x $C_0t_{1/2}$ calculated with respect to the labelled DNA which is in sequence excess. The RNA-DNA hybrids formed in the reaction are analysed by resistance to nuclease S1. The slope of the lines was used to determine the number of transcripts per ng RNA. (Section 10.2).(b) Relationship between [32P]cDNA dot blot hybridisation (*Figure 6*) and the number of transcripts per ng RNA for clones A2, C1 and D2.

suffering from acute non-lymphocytic leukaemia (ANLL). These were screened with [32P]cDNA synthesised using unfractionated mRNA from an ANLL patient as template and with [32P]cDNA complementary to unfractionated mRNA of a patient suffering from chronic lymphocytic leukaemia (CLL). Recombinant clones representing mRNAs common to the two diseases (e.g., B5, C1), and their relative abundance, can be easily identified and distinguished from those apparently specific to ANLL (e.g., D2, D3, D4, D5).

By using a calibration curve, the extent of hybridisation of [32P]cDNA to each filter-bound recombinant can be used to determine the actual number of transcripts of mRNA (47,48). From the dot blot results (*Figure 6*), at least three cDNA clones whose representation in mRNA differs in abundance over a wide range are selected (e.g., clones A2, C1 and D2). Separated single strands of these recombinant DNAs are radiolabelled and hybridised separately in solution to different amounts of mRNA. The concentration of reactants is adjusted such that recombinant DNAs are in sequence excess. The reactions are carried out to about 20 x $C_0t_{1/2}$. The radioactivity in nuclease S1-resistant hybrids is determined and plotted against the amount of mRNA added to the reaction.

As expected, only one of the separated strands of DNA hybridises with the mRNA (*Figure 7a*). The number of transcripts per ng mRNA is then determined from the slope of the line using the relationship:

$$T = \frac{fN}{LS \times 350 \times 10^9}$$

where T is the number of transcripts per ng mRNA, f is the slope of the titration curve (c.p.m./ng mRNA), N is Avogadro's Number (number of molecules/ mol), L is the length of hybrid (i.e., cDNA insert length in nucleotides), S is the specific activity of the labelled cDNA (c.p.m./ng), and 350 is the average molecular weight of a ribonucleotide.

The appropriate values for clones A2, C1 and D2 of *Figure 7a* are:

Clone	Specific activity (S)	Length of insert (L)	Slope (f)	No. of transcripts /ng mRNA (T)
A2	2.0×10^4	800	4.7	5×10^5
C1	1.5×10^4	1000	26.3	3×10^6
D2	1.0×10^4	750	89.4	2×10^7

Using these values of T, a graph is plotted of the number of transcripts against the radioactivity in the hybrids on the dot blots. A linear relationship is obtained (*Figure 7b*). From this graph, the number of transcripts of any other recombinant on the dot blot matrix can easily be obtained.

11. DETERMINATION OF OPTIMAL REACTION CONDITIONS

11.1 Buffer Composition and Temperature

To determine the optimal reaction conditions, prepare replicate dot blots. Hybridise some under different hybridisation conditions, keeping the washing conditions constant, and monitor the effects. Then hybridise other dot blots under optimal hybridisation conditions and vary the washing conditions. For an extensive analysis of the effects of altering conditions on the hybrids formed, the reader is referred to references 3 − 5 and 43. The following is a rough guide.

(i)   Reaction conditions which favour the detection of well-matched hybrids involve high temperatures of hybridisation (65 − 68°C in aqueous solution and 42°C in 50% formamide) combined with washing at high temperatures (5 − 25°C below T_m) and at low salt concentrations (0.1 x SSC).

(ii)   To detect poorly-matched hybrids, filters should be hybridised in solutions containing formamide (20 − 50%) at 35 − 42°C but washed at high salt concentrations at an intermediate temperature (e.g., 2 − 6 x SSC at 40 − 60°C). Again, conditions may have to be determined empirically. It should be remembered that both closely-related and distantly-related sequences will be detected under these conditions.

(iii) To distinguish between closely- and distantly-related members of the same family, conditions must be found which are permissive for some sequences and stringent for others. As we have already seen (Section 5), distant homologies are best detected when the ratio of rate constants for hybridisation of cross-hybridising to self-hybridising species is high, whereas closely-related species are most easily detected when the ratio is low. In practical terms, this means that for distantly-related hybrids low temperatures of incubation are used, whereas for closely-related hybrids high temperatures are best. The time of incubation is very important since the effective discrimination between closely- and distantly-related hybrids is highest with very short times of incubation and deteriorates very rapidly thereafter (Section 5). If short incubations do not give a sufficiently high hybridisation signal, then longer times can be used with excess filter-bound nucleic acid.

11.2 Time Period of Incubation

It is difficult to give a precise time period for hybridising filters. Filter hybridisations tend not to go to completion. As described above, the rate of hybridisation on filters is about 10 times slower than that for solution hybridisation of the same DNAs (9), so it is difficult experimentally to reach the very high C_0t values required for complete hybridisation. Prolonged incubation does not necessarily increase the extent of hybridisation because:

(i) more and more probe reassociates
(ii) at the high temperatures involved, sequences leach off the filter if they are not covalently bound
(iii) the probe is gradually degraded.

Addition of formamide to the hybridisation solution allows lower temperatures to be used and thus incubation times can be extended, but there is still the problem of probe reassociation unless a single-stranded probe is used. Furthermore, steric constraints prevent all the bound sequences being nucleated effectively. Under optimal conditions, no more than 80% of a single sequence probe appears in hybrids (9) and so an even smaller proportion of more complex probes will be hybridised.

In practice, for double-stranded DNA probes, there is no need to proceed for longer than to allow the probe in solution to achieve $1-3$ x $C_0t_{1/2}$. After incubation for 3 x $C_0t_{1/2}$, the amount of probe available for additional hybridisation to sequences on the filter is negligible. The following useful guideline is taken from ref. 46. In 10 ml hybridisation solution, 1 μg denatured double-stranded probe with a complexity of 5 kb will reach $C_0t_{1/2}$ in 2 h. To determine the number of hours (n) needed to achieve $C_0t_{1/2}$ for renaturation of any other probe, the appropriate values can be substituted in the following equation:

$$n = \frac{1}{X} \times \frac{Y}{5} \times \frac{Z}{10} \times 2$$

where X is the weight of the probe added (in μg), Y is its complexity (which for most probes is proportional to the length of the probe in kb) and Z is the volume of the

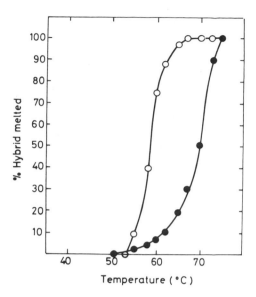

Figure 8. Melting profile of dot hybrids. Replicate filters containing dots of linearised DNA are hybridised to homologous ^{32}P-labelled DNA (○) or RNA (●). Hybridisation, in a formamide-containing solution, is carried out to 20 x $C_o t_{1/2}$. The DNA bound to the filter is present in significant excess. After hybridisation each filter is washed in the same formamide-containing buffer as used for hybridisation, with stepwise increases in temperature. The filters are counted by Cerenkov counting between each washing step to determine the percentage of hybridised [^{32}P]DNA or [^{32}P]RNA which has eluted (melted) at each temperature.

reaction (in ml).

For quantitative dot blots (e.g., *Figure 6*) the time period of incubation, the concentration of probe and the amount of nucleic acid bound to the filter should all be adjusted such that there is low fractional hybridisation of both the probe and filter-bound sequences. This is to ensure that the proportion of the probe hybridising increases linearly with time.

Factors affecting the selection of a time of incubation suitable for discrimination between related sequences are discussed in Section 5.2 [see also Section 11.1 (iii)].

12. MEASUREMENT OF T_m

The DNA or RNA sample is applied in a dot to a filter and hybridised with a labelled DNA or RNA probe as appropriate. It is then washed at the same temperature in a solution of the same composition as was used for the hybridisation. This removes unhybridised and non-specifically-bound probe. For melting of the hybrid, the filter is incubated for 10 – 15 min in a small volume of the same buffer at progressively higher temperatures. The buffers used must be pre-heated to the required temperature. Between melting steps, the filter is counted by Cerenkov counting, making sure that it does not dry out at any stage in order to prevent irreversible binding of the probe. Replicate filters can be hybridised at different criteria or washed in solutions of different ionic strength. *Figure 8* shows the melting temperature profiles to be expected from homologous DNA-DNA and RNA-DNA hybridisations. Note that in formamide-containing solutions RNA-DNA hybrids are more stable than the corresponding DNA-DNA hybrids and so dissociate (melt) at higher temperatures. The exact temperature difference will depend on the base composition of the nucleic acids involved, the size of the probe and the degree of relatedness of the two hybridising species. In the example given in *Figure 8*, the difference in T_m is $11°C$.

For a single nucleic acid species hybridising to itself, the hybrid melts over a very narrow temperature range and the T_m is the same irrespective of the incubation temperature, T_i. However, when the same nucleic acid is probed with a complex mixture of sequences which have varying degrees of relatedness, the T_m profile depends on the reaction conditions. For hybrids formed at low criterion ($T_m -25°C$), the melting profile is broad because both well-matched and poorly-matched hybrids are formed. They melt at different temperatures, so the overall melting profile, which is a composite of the contributions of all the hybrids, will reflect this. At high criterion ($T_m -8°C$), only hybrids with a high degree of homology form so they melt over a very narrow temperature range. The melting profile is also broad when variable length probes are used. This is most apparent at short average lengths of hybridised probe in accordance with the empirical relationship:

$$T_n - T_m = \frac{650}{L}$$

where L is the length of the probe in nucleotides, T_m is the melting temperature of the short hybrid and T_n is the melting temperature of long DNA molecules (17).

The procedure of stepwise melting of hybrids described above for T_m measurement can be extended to investigate the degree of relatedness of different sequences and can be applied to many samples at once. An array of dots on a single filter are hybridised to a labelled probe and the extent of similarity to the probe is evaluated by stepwise melting and autoradiography (3). The more mismatched a hybrid is, the lower the temperature at which it will melt. Thus differences in the intensity of signal of the dots can be interpreted in terms of the degree of relatedness of the different sequences.

13. RE-USE OF FILTERS AND PROBES

13.1 **Filters**

In many cases, re-probing the same filters with a series of different probes yields valuable information. Filters can be re-probed several times — the exact number being dependent on the type of filter and the incubation conditions to which the filter has already been exposed. Nitrocellulose filters which have been exposed to high temperatures for hybridisation and washing can be used about two to four times before falling apart. Filters exposed to the less harsh conditions of hybridisation at lower temprature in the presence of formamide can be used many more times. Nylon filters can be used indefinitely without disintegration and so, because of their superior durability, they are preferred to nitrocellulose for multiple probings.

Before re-probing a filter with a new probe, it is first necessary to strip off the old probe and to monitor that the treatment has been effective. This can be achieved for both DNA and RNA dot blots as follows:

(i)   Transfer the damp filter to a plastic box containing 200 ml of 5 mM Tris-HCl, pH 8.0, 0.1 mM EDTA, 0.05% sodium pyrophosphate, 0.1% Denhardt's solution (see *Table 3*, footnote c for composition) at 65°C. Incubate for 1 h with gentle agitation.

(ii)   Discard the wash buffer and repeat step (i).

(iii)   Check by autoradiography as described in Section 10.1 that the probe has been removed.

An alternative procedure for DNA dot blots only is given below.

(i)   Wash the filter twice, for 10 min each wash, in 50 mM NaOH at room temperature.

(ii)   Wash and neutralise the filter by incubating it (5 min each time) in five changes of TE buffer, pH 7.5 (see *Table 1*, footnote a) at room temperature.

This procedure cannot be carried out successfully for RNA dot blots since the NaOH hydrolyses the filter-bound RNA.

Unfortunately there are several potential problems with re-using filters:

(i)   Loss of sensitivity. Prolonged use of both nylon and nitrocellulose filters leads to gradual reduction in sensitivity through loss of filter-bound nucleic acid (Section 6.1). This is not a problem if the nucleic acid has been covalently bound to the filter matrix.

(ii)   Irreversible binding of the previous probe if the filter was allowed to dry.

(iii)   Incomplete removal of the previous probe even if the filter was kept wet throughout its use. This can give misleading results if single-stranded tails of probe remaining on the filter are complementary to sequences of the new probe. For example, if the first probe is a recombinant DNA (vector and insert) and the second is also a recombinant DNA with a different insert, hybridisation may occur through plasmid sequences and this will obscure the hybridisation of the second probe insert sequences. Therefore, it is recommended that all inserts are excised from vectors before labelling as probes.

13.2 **Probes**

Normally only a small fraction of the probe is used up during hybridisation, so probes can be re-used until they are degraded or have decayed to too low a specific activity. To re-use the probe (now in the hybridisation solution) it must be denatured again by heating to a temperature above its T_m. For aqueous solutions, this can be done by incubating in a boiling water bath for 10 min. For formamide-containing solutions, heat at 70°C for 30 min. The newly-denatured probe can now be added to a second filter which has been pre-hybridised under standard conditions.

14. PROBLEMS

Most hybridisation experiments using filter-bound nucleic acids employ radioactive probes and so problems of only this type of investigation are covered here.

(i)   The autoradiograph of the filter is black all over:
        (a) At some stage during hybridisation or washing, the filter was allowed to

dry. It will probably be necessary to strip the filter (Section 13.1) and re-hybridise.

(b) The probe is 'dirty'. It may be contaminated by traces of agarose. Either re-purify the nucleic acid from which the labelled probe was derived and prepare a new probe, or pass the labelled probe through a nitrocellulose filter which has been pre-treated in 10 x Denhardt's solution or through a mini NACS column (BRL).

(c) An inappropriately low hybridisation and/or washing temperature was used.

(ii)   The autoradiograph of the filter is black in parts:

(a) Part of the filter dried out; see above.

(b) The filter was handled with bare hands. Grease marks from fingers trap probe. Wear disposable plastic gloves in future.

(iii) The autoradiograph has black dots in random locations:

(a) The unincorporated precursors were not completely removed from the probe. See correct procedure in relevant section of Chapter 2.

(b) Air bubbles were not completely removed from the bag during hybridisation. (This may not matter if a shaking water bath is used, but the effect may be quite troublesome if the bag is not agitated.)

(c) Dust or dirt on the filter. Filter all solutions before use in future.

(iv)   The signal is lower than expected:

(a) Was the correct binding procedure used? Nitrocellulose and nylon filters use different binding protocols (see Section 6.1).

(b) The probe was degraded. This is most likely to happen with RNA probes.

(c) The double-stranded probe was not denatured (see Section 8.2.2).

(d) The hybridisation and/or washing conditions were too stringent so that the hybrids either did not form at all, or were dissociated.

(e) The specific activity of the probe was too low.

(f) The hybridisation time was too short.

(g) The filter was not exposed to film for long enough.

(v)   A 'negative' effect is obtained, that is, the background of the autoradiograph is black with clear dots. Too high a concentration of [^{32}P]probe was used.

(v)   The filter fell apart. This is most likely to occur with nitrocellulose filters.

(a) During binding of DNA to the filter, the alkali was not properly neutralised thus making the filter yellowish in colour and very brittle.

(b) After repeated use, the filter becomes brittle despite correct procedures. Prepare new filters.

15. ACKNOWLEDGEMENTS

The authors thank P.Harrison for discussions and The Leukaemia Research Fund of Great Britain for support.

16. REFERENCES

1. Gillespie,D. and Spiegelman,S. (1965) *J. Mol. Biol.*, **12**, 829.
2. Britten,R.J. and Kohne,D.E. (1968) *Science (Wash.)*, **161**, 529.
3. Sim,G.K., Kafatos,F.C., Jones,C.W., Koehler,M.D., Efstratiadis,A. and Maniatis,T. (1979) *Cell*, **18**, 1303.

4. Beltz,G.A., Jacobs,K.A., Eickbush,T.H., Cherbas,P.T. and Kafatos,F.C. (1983) *Methods in Enzymology*, vol. **100**, Wu,R., Grossman,L. and Moldave,U. (eds.), Academic Press, NY, p. 266.
5. Kafatos,F.C., Jones,C.W. and Efstratiadis,A. (1979) *Nucleic Acids Res.*, **7**, 1541.
6. Wetmur,J.G. (1971) *Biopolymers*, **10**, 601.
7. Lee,C.H. and Wetmur,J.G. (1972) *Biopolymers*, **11**, 549.
8. McCarthy,B.J. and McConaughy,B.L. (1968) *Biochem. Genet.*, **2**, 37.
9. Flavell,R.A., Birfelder,E.J., Sanders,J.P. and Borat,P. (1974) *Eur. J. Biochem.*, **47**, 535.
10. Birnsteil,M.L., Sells,B.H. and Purdom,I.F. (1972) *J. Mol. Biol.*, **63**, 21.
11. Flavell,R.A., Borst,P. and Birfelder,E.J. (1974) *Eur. J. Biochem.*, **47**, 545.
12. Wetmur,J.G. and Davidson,N. (1968) *J. Mol. Biol.*, **31**, 349.
13. Marmur,J.G. and Doty,P. (1961) *J. Mol. Biol.*, **3**, 584.
14. Bonner,T.I., Brenner,D.J., Neufield,B.R. and Britten,R.J. (1973) *J. Mol. Biol.*, **81**, 123.
15. Howley,P.M., Israel,M.F., Law,M-F. and Martin,M.A. (1979) *J. Biol. Chem.*, **254**, 4876.
16. Casey,J. and Davidson,N. (1977) *Nucleic Acids Res.*, **4**, 1539.
17. Britten,R.J., Graham,D.E. and Neufield,B.R. (1974) in *Methods in Enzymology*, Vol. **29E**, Grossman,L. and Moldave,K. (eds.), Academic Press, NY, p. 363.
18. Wetmur,J.G. (1975) *Biopolymers*, **14**, 2517.
19. Wahl,G.M., Stern,M. and Stark,G.R. (1979) *Proc. Natl. Acad. Sci. USA*, **76**, 3683.
20. Schildkraut,C. and Lifson,S. (1965) *Biopolymers*, **3**, 195.
21. McConaughy,B.L., Laird,C.D. and McCarthy,B.J. (1969) *Biochemistry (Wash.)*, **8**, 3289.
22. Marmur,J. and Doty,P. (1962) *J. Mol. Biol.*, **5**, 109.
23. Dove,W.F. and Davidson,N. (1962) *J. Mol. Biol.*, **5**, 467.
24. Schmeckpepper,B.J. and Smith,K.D. (1972) *Biochemistry (Wash.)*, **11**, 1319.
25. Hyman,R.W., Brunovskis,I. and Summers,W.C. (1973) *J. Mol. Biol.*, **77**, 189.
26. Yang,R.C., Young,A. and Wu,R. (1980) *J. Virol.*, **34**, 416.
27. Haas,M., Vogt,M. and Dulbecco,R. (1972) *Proc. Natl. Acad. Sci. USA*, **69**, 2169.
28. Church,G.M. and Gilbert,W. (1984) *Proc. Natl. Acad. Sci. USA*, **81**, 1991.
29. Southern,E.M. (1975) *J. Mol. Biol.*, **98**, 503.
30. Nagamine,Y., Sentenac,A. and Fromageot,P. (1980) *Nucleic Acids Res.*, **8**, 2453.
31. Parnes,J.R., Velan,B., Felsenfeld,A., Ramanathan,L., Ferrini,U., Apella,E. and Seidman,J.G. (1981) *Proc. Natl. Acad. Sci. USA*, **78**, 2253.
32. Thomas,P. (1983) in *Methods in Enzymology*, Vol. **100**, Wu,R., Grossman,L. and Moldave,U. (eds.), Academic Press, NY p. 255.
33. McMaster,G.K. and Carmichael,G.G. (1977) *Proc. Natl. Acad. Sci. USA*, **74**, 4835.
34. Bailey,J.M. and Davidson,N. (1976) *Anal. Biochem.*, **70**, 75.
35. Lehrach,H., Diamond,J., Wozney,J.M. and Boedtker,H. (1977) *Biochemistry (Wash.)*, **16**, 4743.
36. Bantle,J.A., Maxwell,I.H. and Hahn,W.E. (1976) *Anal. Biochem.*, **72**, 413.
37. Thomas,P. (1980) *Proc. Natl. Acad. Sci. USA*, **77**, 5201.
38. Renz,M. and Kurz,C. (1984) *Nucleic Acids Res.*, **12**, 3435.
39. Langer,P.R., Waldrop,A.A. and Ward,D.C. (1981) *Proc. Natl. Acad. Sci. USA*, **78**, 6633.
40. Leary,J.J., Brigati,D.J. and Ward,D.C. (1983) *Proc. Natl. Acad. Sci. USA*, **80**, 4045.
41. Denhardt,D.T. (1966) *Biochem. Biophys. Res. Commun.*, **23**, 641.
42. Singh,L. and Jones,K.W. (1984) *Nucleic Acids Res.*, **12**, 5627.
43. Meinkoth,J. and Wahl,G. (1984) *Anal. Biochem.*, **138**, 267.
44. Williams,J.G. and Lloyd,M.M. (1979) *J. Mol. Biol.*, **129**, 19.
45. Dworkin,M. and Dawid,I.B. (1980) *Dev. Biol.*, **76**, 435.
46. Maniatis,T., Fritsch,E.F. and Sambrook,J. (1982) *Molecular Cloning, A Laboratory Manual*, Cold Spring Harbor Laboratory Press, NY.
47. Lasky,L.A., Lev,Z., Xin,J-H., Britten,R.J. and Davidson,E.H. (1980) *Proc. Natl. Acad. Sci. USA*, **77**, 5317.
48. Xin,J-H., Brandhorst,B.P., Britten,R.J. and Davidson,E.H. (1982) *Dev. Biol.*, **89**, 527.

111

CHAPTER 5

Hybridisation in the Analysis of Recombinant DNA

PHILIP J. MASON and JEFFREY G. WILLIAMS

1. INTRODUCTION

The screening methods described in this chapter are used to identify and purify a particular recombinant DNA molecule. It is convenient to divide these methods into primary and secondary screening techniques.

In *primary screening*, colonies of bacteria transformed with recombinant plasmids or bacteriophage plaques are replicated *in situ* onto a nitrocellulose filter. The DNA is then liberated, denatured, and bound to the filter. When a specific probe such as a cloned DNA fragment is available, it is hybridised to the immobilised DNA and used to identify recombinants containing complementary sequences. If a specific probe is not available, then a number of strategies may be used to prepare a probe enriched in the required sequence. In this case primary screening is used to identify, from the very large number ($>10^6$) of recombinants that can be screened by *in situ* hybridisation, a small number ($<10^2$) of potential positives that can then be subjected to secondary screening.

In *secondary screening*, DNA preparations from individual recombinants (or pooled DNA from a small number of recombinants) are individually bound to an inert support such as nitrocellulose. This filter-bound DNA is then used to hybridise complementary mRNA sequences from total cellular RNA. The selected mRNA is eluted from the DNA, translated *in vitro* and the encoded polypeptide identified. This procedure is called hybrid selection. This chapter also describes the DNA dot-blot technique which can be useful both in secondary screening and in the analysis of a cloned DNA. Finally, the analytical procedure described in this chapter is Southern blotting by which individual fragments of DNA are separated by gel electrophoresis, transferred to an inert support such as nitrocellulose and then analysed by hybridisation with a suitable radioactively-labelled probe.

2. PRIMARY SCREENING TECHNIQUES

2.1 Replication, Storage and Immobilisation of Recombinant DNA Libraries

2.1.1 *General Comments*

In all of the techniques described in this chapter, DNA is immobilised onto an inert support. Apart from the use of chemically activated paper in hybrid selection, nitrocellulose is used in all the procedures. Single-stranded DNA binds to nitrocellulose in solutions containing a high salt concentration. The binding is reversible but, after drying the filters at high temperature for several hours, the DNA becomes tightly bound although it remains available for hybridisation. Nitrocellulose must be handled very carefully at all times as it is a very fragile material. It is also very inflammable and so great care must be taken not to overheat when drying or baking these filters. Overheating also greatly increases the fragility of nitrocellulose. Recently, a number of alternative filter materials which are more robust have become commercially available. It may be advisable to use one of these if the filter is to be hybridised sequentially with a number of different probes. With care, however, it is possible to use a nitrocellulose filter at least two or three times before it is transformed into a 'jigsaw puzzle'. We have used nitrocellulose filters of pore size 0.45 μm from either Schleicher and Schüll or Sartorius.

The details of the procedure required to immobilise DNA from bacterial plasmids or cosmids differ slightly from those used for phage λ or M13 libraries. Therefore the following section gives the basic method for screening bacterial colonies and then Sections 2.1.3 and 2.1.4 list the differences when using bacteriophage vectors.

2.1.2 Bacterial Colonies

Basic protocol. The protocol used in the authors' laboratory is based on those of Grunstein and Hogness (1) and Hanahan and Meselson (2). It is useful whether screening hundreds of thousands of bacteria for a rare cDNA clone or a few dozen for a particular subclone. In outline, the bacteria are grown on agar plates and replicas of the plates are then made onto nitrocellulose filters. The colonies are grown on the filters and lysed, the DNA is then denatured *in situ* and bound to the filters. This 'library' (or 'bank') is then ready to be screened by hybridisation with a radioactive probe. The detailed protocol is as follows.

(i) Prepare L agar plates (10 g Tryptone, 5 g yeast extract, 15 g agar, 5 g NaCl per litre) containing the appropriate antibiotic and plate out the bacteria on these. The colony density may be as high as 200 per square centimetre. Incubate the plates at 37°C until the colonies are just visible to the naked eye.

(ii) Cut nitrocellulose sheets into pieces that fit comfortably into the Petri dishes. *Always* wear gloves when handling nitrocellulose or background 'hybridisation' will be generated where the filter contacts the skin. If the library is to be plated onto more than one plate, number the nitrocellulose filters with a soft pencil or ball-point pen.

(iii) Lay each dry nitrocellulose filter carefully onto the corresponding agar surface. The best way to do this is to grip the filter at two opposing edges with flat-bladed forceps (e.g., Millipore forceps) and hold it above the dish, bowing slightly down in the middle. Lower the filter until it contacts the agar at one point. Slowly lower it further and watch the filter wet evenly all over starting from the point of initial contact. If the nitrocellulose you are using has a shiny side and a matt side, always put the shiny side towards the agar.

(iv) Fill a small syringe with black indelible ink and blot the end of the syringe needle until the ink is flowing slowly and evenly. Now prick through each filter into the agar in an asymmetric pattern in order to 'key' the replica filter to the original plate.

(v)   Carefully peel off the filter and lay it, colony side up, on a fresh agar plate containing the appropriate antibiotic. Try not to trap any air bubbles between the filter and the agar.

(vi)   Make at least one more replica of the original master plates in the same way, remembering to key all filters to the master plate.

(vii)   Incubate the nitrocellulose replicas and the original agar plate at 37°C until the colonies are about $1-2$ mm in diameter but still clearly distinct. This will take longer for the second and subsequent replicas than for the first one. Then store the master plate at 4°C until screening (Section 2.2) is completed.

(viii) If using a vector such as pBR322, which can be amplified using chloramphenicol, peel off the filters and place them colony side up on agar plates containing chloramphenicol at 100 μg/ml and incubate the plates overnight at 37°C. This step is not worthwhile when using vectors such as pAT153 (ref. 3) or the pUC series (4) which accumulate to high copy number in the absence of chloramphenicol.

(ix)   Using forceps, peel each filter off its plate and lay it, colony side up, on a stack of three sheets of 3MM paper (Whatman) saturated with denaturation solution (1.5 M NaCl, 0.5 M NaOH). This is conveniently done in plastic trays. Excess solution should be poured out of the tray at all steps in the procedure so that there are no surface pools of liquid.

(x)   Transfer the filter, keeping it horizontal, to a stack of 3MM paper saturated with neutralisation buffer (1.5 M NaCl, 1 M Tris-HCl, pH 7.5). After 5 min, transfer the filter to paper saturated with 4 x SET buffer (20 x SET buffer stock has the composition; 3 M NaCl, 20 mM EDTA, 0.4 M Tris-HCl, pH 7.8)

(xi)   Next place the filter on dry 3MM paper and leave at room temperature or at 37°C until it is dry $(15-30$ min).

(xii)   Sandwich the filters between sheets of 3MM paper. Either tape the 3MM sheets together at the sides or put the stack of filter sandwiches between two glass plates. Bake the filters in an oven at 80°C for 2 h.

Long-term storage of plasmid libraries. It may be desirable to store a library of recombinant plasmids as an array of bacterial colonies on nitrocellulose filters so that they can be screened with several probes over a period of months. This is most conveniently done if the library is plated out initially on nitrocellulose, rather than on agar plates as described above. Small amounts of detergent will inhibit bacterial growth and so a special grade of nitrocellulose (Millipore HATF) should be used for plating. The protocol is described in *Table 1*.

2.1.3 *Bacteriophage* λ

The procedure for replication of a bacteriophage λ library onto nitrocellulose filters differs from the protocol for bacterial colonies (Section 2.1.2). The procedure, basically as described by Benton and Davis (5), is as follows.

(i)   Plate the phage λ library in top agar or top agarose (5). (Top agar contains 8 g agar, 10 g Tryptone, 5 g yeast extract and 5 g NaCl per litre. Top agarose contains 2% agarose instead of agar.) The plaque density may be as high as semi-confluent, that is, when plaques touch their neighbours but individual plaques are still discernible.

Table 1. Procedure for Long-term Storage of Plasmid Libraries on Nitrocellulose Filters.

1.	Cut detergent-free nitrocellulose filters (Millipore HATF) to the required size and sterilise them by autoclaving in a sealed autoclavable plastic bag.
2.	Plate out the library onto the sterile filters on L-agar plates[a] that contain the appropriate antibiotic and 5% glycerol. These will be the master filters.
3.	Incubate the plates at 37°C. When the colonies are just visible, peel off each filter and place it (colony side uppermost) on a sheet of 3MM paper on a flat surface.
4.	Wet a second filter by briefly laying it on a plate containing 5% glycerol. Now place this second filter on top of the first and sandwich them together by applying pressure with a replica plating disc. Key the two filters together using indelible ink [Section 2.1.2, step (iv)].
5.	Wrap the sandwich first in 3MM paper and then in foil. Store frozen at −70°C.
6.	To screen the colonies, separate the two filters (when still frozen) and incubate each on an L-agar plate containing the appropriate antibiotic for a few hours at 37°C. The agar used to support the master filter should also contain 5% glycerol.
7.	Incubate the filters until colonies are visible. Sandwich and freeze the master filter as before for long-term storage (steps 4 and 5). Place the second (replica) filter (colony side uppermost) on 3MM paper. Lay a wetted piece of nitrocellulose onto the filter, press the two filters together, and then separate them as before (steps 4−6). Several (up to 5 or 6) copies of the replica filters can be made in this way. The replica filters are now ready to be prepared for hybridisation as described in Section 2.1.2 steps (vii)−(xii).

[a]See Section 2.1.2, step (i).

(ii)   Put the plates at 4°C for 30 min. This step helps to prevent the top agar or agarose sticking to the nitrocellulose filter.

(iii)   Place a filter onto the plate, key it to the agar plate [Section 2.1.2, step (iv)] and leave it to absorb for one minute. Peel off the filter.

(iv)   Immediately carry out the lysis/denaturation step [Section 2.1.2, step (ix)].

(v)   Place a second filter onto the plate, key it, and allow it to absorb for about 3 min. Take this filter through the lysis/denaturation step [Section 2.1.2, step (ix)] also.

(vi)   The procedure for the immobilisation of the DNA on the nitrocellulose is the same as for bacterial colonies. [Section 2.1.2, Steps (x)−(xii)]. Store the agar plates at 4°C until the screening (Section 2.2) is completed.

2.1.4 *Bacteriophage M13*

The male-specific phage M13 has been adapted for use as a cloning vector (6). The intracellular replicative form is a double-stranded circle, whilst the phage particle contains a single-stranded DNA molecule. The procedure for replication of an M13 phage library on nitrocellulose filters is basically the same as for phage λ (Section 2.1.3) except that we have obtained the best results if the filters are baked, untreated, immediately after the plaque lift [Section 2.1.3, step (iii)], that is, with no lysis or denaturation steps. Obviously plaque screening will only be successful when the probe to be used contains sequences complementary to the strand of the cloned DNA that is packaged in the phage.

2.1.5 *The Micro-titre Tray Method*

This procedure is very useful when screening relatively low numbers of bacterial colonies (<2000). The colonies are picked from the original plate into the wells of standard (96-well) micro-titre trays, grown to saturation and stored at −20°C or −70°C after

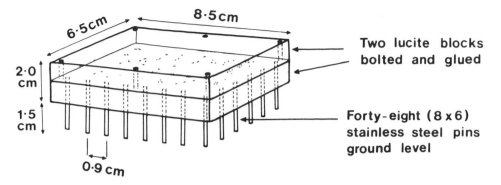

Figure 1. A replica-plating device for microtitre trays. This apparatus is designed for use with a standard (96 well) microtitre tray. It is constructed from two Perspex blocks and 48 stainless steel pins or nails. The pins are inserted into holes in the lower of two blocks and held in place by the upper block. It is important that the heads of all the pins be ground to a uniform depth below the lower block so that they all make an equal contact with the nitrocellulose filter during replication.

the addition of a freezing medium (such as glycerol or DMSO). The bacteria remain viable under these conditions for many years. Therefore the technique is particularly useful as a means of storing a 'bank' of cDNA clones for future use. The bank is easily replicated into and from micro-titre trays using a device similar to that shown in *Figure 1*. Screening for a particular insert is a very simple process because the required clone can be identified in just one round of selection. It is also an excellent method for performing 'differential screening', that is, screening a set of clones with more than one probe in order to identify cloned sequences which are expressed at different levels in different tissues or at different stages of development. Finally it has the advantage that colonies are individually isolated from the primary transformation plates. Thus cloned inserts which have a deleterious effect in the bacterial host do not become under-represented as they do in high density procedures which involve several rounds of screening. With suitable modifications, the same method can be used for bacteriophage colonies (7). The major disadvantages are that it is much more time consuming and more wasteful of nitrocellulose and storage space than the high density screening methods described in Section 2.1.2. This limits its usefulness to screening a cDNA clone bank for sequences in the medium to high mRNA abundance classes (i.e., >0.5% of the mRNA population).

The detailed procedure for the micro-titre tray method is as follows.

(i)   Add to the wells of a standard (96-well) microtitre tray (e.g., Flow Laboratories Cat. No. 76-003-05) 150 μl of bacterial growth medium containing the required amount of the antibiotic suitable for selection of transformed colonies. (If a suitable multiple delivery system is not available, two drops from a 2 ml graduated pipette gives approximately the correct volume).

(ii)   Transfer the bacterial colonies into individual wells using sterile wooden toothpicks. It is advisable to pick *all* sizes of colony rather than to be selective. Slow growth of a transformant may be the result of a deleterious effect of the insert. It is helpful to leave the toothpicks in the wells until the tray has been completely filled as a guide to which wells contain a colony. Leave one well

117

in each half of every tray uninoculated as a sterility check and as an orientation guide in subsequent manipulations. The dimensions of the standard 96-well microtitre trays are such that the right and left halves of the tray can be conveniently replicated separately onto nitrocellulose filters lying on 9 cm diameter agar plates.

(iii)   Remove the toothpicks carefully so as not to cross-contaminate the wells.

(iv)   Stick a strip of adhesive film (e.g., Falcon Laboratories Cat. No. 3044F) onto each tray to separate the wells and prevent contamination. Place at 37°C overnight to allow the bacteria to grow.

(v)   Using a template (8.5 cm diameter) cut sufficient nitrocellulose circles for the number of colonies to be replicated (i.e., twice the number of complete trays). Label and orientate the filters with a soft lead pencil then lay them onto agar plates containing antibiotic.

(vi)   Stand a replica plating device (*Figure 1*) in a Petri dish (9 cm diameter) containing 70% ethanol. Remove the device and flame off the ethanol *using a minimal amount of heat*. Wait until the device has cooled sufficiently to prevent overheating of the bacteria. (It is advisable to wear sterile disposable gloves while performing these manipulations. The temperature of the metal prongs can therefore be checked by touching them onto the back of the gloved hand.)

(vii)   Place the replica plating device into the wells in the appropriate half of the tray. Then carefully lift it just above the surface of the liquid to allow drops to fall off and lay it down gently onto the surface of the nitrocellulose filters.

(viii)   Wait 5 sec then remove the replica plating device and repeat the sterilisation procedure [step (vi)].

(ix)   Incubate the replica filters overnight at 37°C to allow the bacteria to grow.

(x)   Perform alkali lysis and immobilisation of DNA prior to hybridisation exactly as described in Section 2.1.2, steps (ix)−(xii).

There are two alternative storage methods for the bacteria in the microtitre trays, depending on whether they are to be stored for only a short time or for a longer period (>1 year). The short-term storage procedure is useful because it is not necessary to thaw out the contents of the wells before replication. However, for a particularly valuable bank which will require frequent re-probing, it is advisable to make two copies; one for short-term storage at −20°C which can be used for replica plating and another master copy which is kept frozen at −70°C for long-term storage. Both procedures are given in *Table 2*.

Table 2. Procedures for Storage of Plasmid Libraries in Micro-titre Trays.

Short term storage (<1 year)

1.   Add to each well of the micro-titre tray 150 µl of sterile glycerol (2 drops from a 5 ml wide-bore pipette).

2.   Seal the trays using adhesive film taking care not to cross-contaminate the wells.

3.   Agitate on a micro-titre plate shaker (e.g., Dynatech Ltd) for 5 min and then stack the trays in a −20°C freezer.

Long term storage (>1 year)

1.   Add to each well of the micro-titre tray, 150 µl of 14% DMSO in nutrient medium.

2.   Seal the trays using adhesive film as above.

3.   Agitate the trays gently by hand and then stack them in a −70°C freezer.

2.2 **Hybridisation of Filter-bound DNA**

2.2.1 *Standard Procedure Using Polynucleotide Probes*

Hybridisation of radioactively-labelled nucleic acids to filter-bound DNA is a three-stage process. The filter is first pre-hybridised in a solution containing heterologous single-stranded nucleic acid, SDS and 'Denhardt's solution' (8). The purpose of this step is to saturate binding sites on the nitrocellulose that would otherwise lead to an unacceptable background. The filter is then incubated in the same solution containing the radioactively-labelled probe for sufficient time to allow hybridisation to take place. The filters are then washed under conditions of temperature and salt concentration such that only specific hybrids are stable. The location of the hybrid molecules is then determined by autoradiography. The exact details of the hybridisation protocol will depend on the aims of the particular experiment and the probe being used. The basic protocol (*Table 3*) is suitable when using nick-translated double-stranded DNA to probe a set of filters prepared as described in Section 2.1. It can also be used, with little modification, in other screening experiments and in hybridising Southern blots (Section 3.3). As an alternative to the aqueous conditions used in this protocol, the pre-hybridisation and hybridisation steps can be carried out in 50% deionised formamide, 5 x SET buffer, and the temperature lowered to 42°C. The rest of the procedure and the other constituents of the buffers are as described. The lower temperature used in formamide hybridisations prolongs the life of the filters allowing them to be used many times. In general, however, the authors have obtained the best results using aqueous conditions.

In some experiments close attention will have to be paid to the stringency of hybridisation. For example, when the probe is related to a number of different sequences in the filter-bound DNA, these may all give a signal at low stringency but only very closely related, or identical, sequences will give a signal at high stringency. Similarly, if a probe from one organism is used to identify a related sequence in DNA from a different organism, lower stringency conditions will have to be used.

In order to identify sequences very closely related to the probe, perform the hybridisation as described (*Table 3*) but include a final wash at high stringency in 0.1 x SET buffer 0.1% SDS at 68°C for about 20 min. Expose the filter to X-ray film without allowing the filter to dry completely. If necessary perform more washes at higher stringency by reducing the salt concentration further (0.05 x SET, 0.1% SDS). To identify sequences with limited homology to the probe, perform the hybridisation as described (*Table 3*) but reduce the temperature of hybridisation and washing. A final wash in 2 x SET buffer at 50°C, for example, should allow identification of sequences with only about 75% homology with the probe. However, the exact relationship between stringency and sequence homology depends on base composition, the length of the probe and of the homologous regions. Hence it is difficult to determine other than empirically. When proposing to screen a DNA library at low stringency, it is often helpful to establish the highest stringency at which the hybrids are stable by performing the appropriate Southern or Northern blot (Section 3.3 and Chapter 6, Section 2, respectively). These conditions may then safely be used in screening the library.

Finally, in certain cases it may be useful to enrich the probe in specific sequences prior to use. This is discussed in Chapter 2, Section 3.4.2.

Table 3. The Standard Hybridisation Procedure using Polynucleotide Probes.

1.	Prepare the following stock solutions: 20 x SET buffer (3 M NaCl, 20 mM EDTA, 0.4 M Tris-HCl, pH 7.8) 10% SDS 5% sodium pyrophosphate Pre-hybridisation solution (4 x SET buffer, 10 x Denhardt's solution[a], 0.1% SDS, 0.1% sodium pyrophosphate, 50 μg/ml denatured salmon sperm DNA[b]).
2.	Wet the filters one at a time by carefully floating them on the surface of a solution of 4 x SET until wetted evenly. Then soak them in 4 x SET buffer for 15 min at room temperature.
3.	Place the damp filters in a polythene freezer bag. Seal the bag along three sides and examine for possible leaks. Several filters may be stacked together in a single bag.
4.	Add at least 1 ml of pre-hybridisation solution per 10 cm² of filter. To avoid placing too many filters or too little liquid in each bag, check that the filters are able to move freely in the bag and do not stick to each other when the bag is moved.
5.	The next operation is to remove all air bubbles. To do this, seal the bag completely, then snip off one corner and displace the bubbles out of that corner by lying the bag on a flat surface and running a ruler along the bag. Removing bubbles is especially important if it is not possible to agitate the bag during pre-hybridisation. Then completely seal the bag and place at 68°C with gentle agitation for at least 2 h and up to 16 h. This can be done in a shaking water bath or on a shaking table inside an oven.
6.	If a double-stranded DNA probe is to be used, it must be denatured prior to hybridisation. Single-stranded DNA (e.g., from M13 phage) or mRNA probes do not require denaturation. To denature a double-stranded DNA probe, heat it to 100°C in water or TE buffer (10 mM Tris-HCl, pH 7.5, 1 mM EDTA) for 3 min and then quickly cool on ice. Alternatively (and more safely) denature the probe with alkali. To do this, add 1/10th original volume of 3 M NaOH and leave at room temperature for 5 min. Then neutralise with 1/5th original volume of 1 M Tris-HCl, pH 7.0, followed by 1/10th original volume of 3 M HCl.
7.	Open the bag by cutting off one corner. If the volume of pre-hybridisation buffer is larger than that required during hybridisation, empty out all of the buffer and pour in the required volume of hybridisation solution (pre-hybridisation solution containing the labelled probe). Alternatively mix the denatured probe with 2 or 3 volumes of pre-hybridisation buffer and add to the pre-hybridisation buffer in the bag using a disposable pipette. In general, the volume of liquid in the bag should be as small as possible whilst still permitting the filters to move about independently. No part of the filter must be allowed to dry during the hybridisation or background will be generated. The amount of probe to be used will depend on the particular experiment and the amount of probe available. As a general guide add 10^5-10^6 c.p.m./ml of nick-translated probe at a specific activity of 5 x 10^7-5 x 10^8 c.p.m./μg. Between 10^5 and 10^6 c.p.m./ml is also a good range for cDNA and mRNA probes.
8.	Now seal the bag, avoiding air bubbles and trying not to spill any radioactive material out of the bag. This step should be performed on a piece of Benchkote (Whatman) in a tray.
9.	Place the bag, on a plastic tray lined with Benchkote, in an oven. Alternatively seal the bag inside a second bag and incubate the bags in an oven or in a water bath at 68°C, with shaking, for the required period. This is normally for 16–24 h, by which time the nick-translated probe will have re-annealed. Only when the concentration of the hybridising species in the probe solution is low (e.g., when trying to obtain hybridisation to a low abundance mRNA species or its cDNA copy in a complex population) will any advantage be gained by hybridising for longer periods of time.
10.	Open the bag at one corner and pour the contents into a plastic tube or flask. This operation is best performed over a sink designated for radioactive disposal. Mark the tube or flask with radioactive tape and place it behind a suitable Perspex or lead shield.
11.	Open the bag over the sink and remove the filters. Place them in a plastic or glass tray containing enough 4 x SET buffer, 0.1% SDS, to immerse the filters completely. Swirl them around in the solution for about 1 min and then pour the buffer away. It is important not to let the filters dry out during this stage or an increased background in the final autoradiograph (step 15) will be generated.

12.   Wash the filters according to the following schedule in a shaking water bath or in an oven. Bring
       all buffers to the correct temperature ahead of time using a microwave oven or a water bath:

   Three times for 20 min in 3 x SET, 0.1% SDS, 0.1% sodium pyrophosphate at 68°C
   Twice for 20 min in 1 x SET, 0.1% SDS, 0.1% sodium pyrophosphate at 68°C
   Once for 20 min in 0.1 x SET, 0.1% SDS, 0.1% sodium pyrophosphate at 68°C (optional
      high stringency wash)
   Once for 20 min in 4 x SET at room temperature.

   In each case, use as much wash solution as is practicable but at least 5 ml per 10 cm² of filter.

13.   Allow the filters to dry on a sheet of 3MM paper at room temperature or at 37°C. (Do *not* dry
       the filters if they are to be re-washed or re-screened).
14.   Mark the filters at the keying marks [Section 2.1.2, step (iv)] with radioactive ink (black ink con-
       taining waste ^{32}P at ~ 10^6 c.p.m./ml).
15.   Wrap the filters in cling film and expose them to X-ray film. If intensification is required, pre-flash
       the film (9) and place it between the filters and an intensifying screen with the flashed side of the
       film next to the screen. Expose the film in a cassette at −70°C and develop according to the manufac-
       turer's instructions.

---

[a]100 x Denhardt's solution contains 2% bovine serum albumin (Sigma, Fraction V), 2% Ficoll, 2%
polyvinylpyrrolidone. Stored at −20°C.
[b]Denatured salmon sperm DNA (1 mg/ml): Dissolve salmon sperm DNA (Sigma) in water at 1 mg/ml,
sonicate to a length of 200−500 bp, stand in a boiling water bath for 20 min, add water if necessary to
make a final solution of 1 mg/ml. Store at −20°C.

2.2.2 Re-use of Filters and Probes

Filters. It is often useful to hybridise the same filter sequentially to a number of different probes. In this case the filter must not be allowed to dry completely or the probe
may become so tightly bound that it cannot be removed. Therefore, if you are going
to re-use the filters, at the end of the washing procedure (*Table 3*, step 12) and while
the filters are still damp, seal them in thin plastic film ('cling-film') and expose them
to X-ray film. Having obtained a suitable X-ray image, wash off the original probe
as follows.

(i)   Immerse the filter in a solution of 30 mM NaOH and incubate at room temperature
        for 10 min with gentle agitation.
(ii)   Now immerse the filter in 10 mM Tris-HCl (pH 7.5), 1 mM EDTA and incubate
        at room temperature for 10 min with gentle agitation.
(iii)   Repeat step (ii) twice more.
(iv)   To verify that the original probe has been efficiently removed in steps (i)−(iii),
        wrap the damp filter in cling-film and expose it to X-ray film (*Table 3*, step 15).
        Assuming that this shows the probe has been removed, the filter can then be
        hybridised to the next probe of choice.

Probes. In most hybridisations, an insignificant amount of the probe in the solution
actually hybridises to the filter so probes can be used repeatedly until their specific
radioactivity has decayed to such a low level or they have degraded to too small a size
to be useful. With double-stranded probes, remember to denature the probe before *each*
hybridisation. The method used will depend on the composition of the medium used
in the original hybridisation. Thus, if the probe is in 4 x SET buffer, heat the solution

to 100°C and maintain this temperature for 10 min. Alternatively, if the probe is in 5 x SET buffer, 50% formamide, heat the probe to 70°C for 10 min. In both cases, these heating steps should be performed by placing the probe in a suitable container and heating in a water bath. Monitor the temperature using a clean thermometer to ensure that the T_m has been exceeded.

2.2.4 *Screening with Oligonucleotide Probes*

When some of the amino acid sequence of a gene product is known, it may be possible to predict the nucleotide sequence of the corresponding portion of the gene and then synthesise an oligonucleotide that can be used to identify clones containing the desired sequences (10,11). However, due to redundancy in the genetic code it is not usually possible to predict a unique nucleotide sequence from the known amino acid sequence and so a mixture of oligonucleotides will have to be synthesised containing all the alternative coding sequences. The number of possible alternatives can be reduced by choosing a part of the sequence rich in tryptophan and methionine residues, which have single codons, and/or residues which have only two codons each (Glu, Gln, Asp, Asn, Phe, Tyr, His, Cys, Lys). The oligonucleotide mixture is normally made in one synthesis by adding mixtures of the appropriate nucleotides at points in the synthesis where there is redundancy. Ideally, the oligonucleotides should be 17 bases long, that is, containing five codons and the first two unambiguous bases of the sixth codon. This is because 17 bases is enough to obtain a unique sequence whereas increasing the length beyond 17 bases decreases the concentration of the correct sequence in the mixture (because of more redundancy). The example shown below is the mixture of 48 seventeen-mers used by Woods *et al.* (12) to identify cDNA clones for the human complement protein factor B.

Amino acid sequence	Asn	Tyr	Asn	Ile	Asn	Gly
Synthetic oligonucleotide	$^{3'}TT^A_G$	AT^A_G	TT^A_G	$TA^A_{G\atop T}$	TT^A_G	$CC^{5'}$
mixture (17−mer)						

The overall strategy for screening with oligonucleotide probes is given in *Table 4*.

Oligonucleotide screening is also used to identify the products of oligonucleotide-directed mutagenesis (14,15). In this case the hybridisation and washing procedures must distinguish between sequences having perfect complementarity with the probe and those with one or a few mismatches. This is done by hybridising at a temperature well below the T_d, the temperature at which a perfectly matched hybrid is half-dissociated (see *Table 4*), and then gradually increasing the washing temperature (in 6 x SSC, 0.1% SDS) until imperfect hybrids are removed. An example of this technique is shown in *Figure 2*.

2.3 Interpretation of Results and Problem Solving

2.3.1 *Identification of Positive Signals*

In interpreting the developed X-ray films from a screening experiment, it is important to distinguish real signals resulting from hybridisation of the probe to cloned DNA from background artifacts. For this reason it is advisable to hybridise duplicate filters, that is, two filters taken from the same master plate and hybridised with the same probe. Only signals that duplicate should be regarded as authentic positives (see *Figure 3*). The appearance of authentic signals is often diagnostic: they have the size and shape

Table 4. Screening with Oligonucleotide Probes.

1. Label the oligonucleotides to high specific activity with $[\gamma\text{-}^{32}P]ATP$ and T4 polynucleotide kinase (see Chapter 2).

2. Ideally the hybridisation reaction with the oligonucleotide probe should be carried out at a temperature at which perfect hybrids are stable but mismatched hybrids are unstable. This is achieved by hybridisation at about 5°C below the temperature (T_d) at which a perfectly-matched hybrid will be half-dissociated. The following relationship (13) gives a fairly good estimate of T_d:

$$T_d = 4°C \text{ per GC base pair} + 2°C \text{ per AT base pair}$$

   For a mixed oligonucleotide, a hybridisation temperature 5°C below the lowest T_d $(T_{d_{min}})$ should be used.

3. Pre-hybridise the filters in the pre-hybridisation buffer (6 x SSC[a], 10 x Denhardt's solution[b], 50 μg/ml denatured salmon sperm DNA[b]) at the hybridisation temperature for at least 1 h. The background radioactivity retained on filters is more intense when using oligonucleotide probes; hence longer pre-hybridisation at 65°C with several changes of buffer may be necessary when screening bacterial colonies.

4. Hybridise in pre-hybridisation buffer [see step 3] containing 0.2 – 1 ng/ml of each oligonucleotide in the mixture, for at least 2 h.

5. Wash the filters in 6 x SSC, 0.1% SDS with several changes of buffer at the hybridisation temperature. Then wash at the T_d or $T_{d_{min}}$ for 2 min. Mismatched hybrids should dissociate much faster than perfectly-matched hybrids at this temperature.

6. Expose the filters to X-ray film as described in *Table 3*, steps 13 – 15.

[a]The composition of standard saline citrate (SSC) buffer is 0.15 M NaCl, 0.015 M trisodium citrate.
[b]See *Table 3* footnotes for details of preparation.

of the colonies or plaques and often have comet-like tails resulting from smearing at the lifting step. Background signals usually do not have a regular size and never have the characteristic tail. It is especially important to use duplicate filters when performing low stringency hybridisations and when using oligonucleotide probes since the background is higher in these cases.

2.3.2 Reduction of Background

The presence of 10 x Denhardt's solution and denatured DNA in the pre-hybridisation and hybridisation media (*Tables 3* and *4*) reduces background on the filter resulting from non-specific hybridisation. Occasionally, however, other measures need to be taken to achieve a satisfactory signal-to-noise ratio.

(i) When screening bacterial colonies, debris from the lysed bacteria can cause background. This can be reduced by pre-hybridisation, over a period of 24 h, with several changes of buffer. Alternatively the bacterial debris can be removed by gently wetting the filters with pre-hybridisation solution and rubbing them gently with a paper towel.

(ii) If the probe is particularly rich in A and T residues [e.g., poly(A)+ mRNA or cDNA copied from it], the background can be reduced by including 10 μg/ml poly(A) in the pre-hybridisation and hybridisation solutions (*Tables 3* and *4*).

(iii) If the probe is from a cDNA clone constructed by poly(dG)/(dC) tailing (17), the inclusion of 10 μg/ml of poly(dC) will reduce background.

(iv) Some background problems can be solved by incubating the hybridisation solution with blank sheets of nitrocellulose for a few hours prior to hybridisation

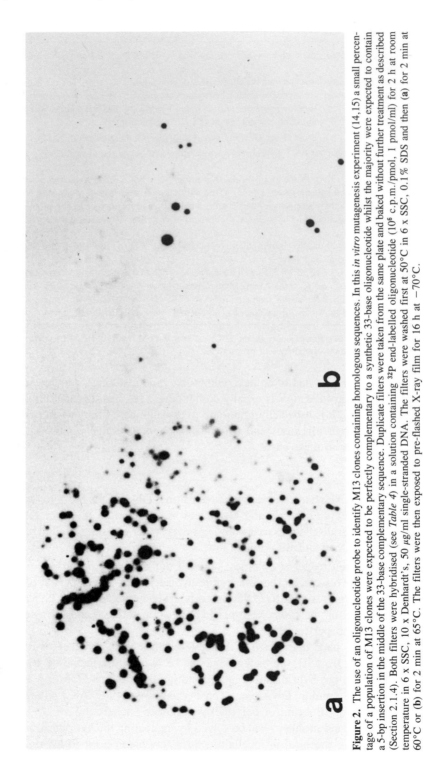

Figure 2. The use of an oligonucleotide probe to identify M13 clones containing homologous sequences. In this *in vitro* mutagenesis experiment (14,15) a small percentage of a population of M13 clones were expected to be perfectly complementary to a synthetic 33-base oligonucleotide whilst the majority were expected to contain a 5-bp insertion in the middle of the 33-base complementary sequence. Duplicate filters were taken from the same plate and baked without further treatment as described (Section 2.1.4). Both filters were hybridised (see *Table 4*) in a solution containing ^{32}P end-labelled oligonucleotide (10^6 c.p.m./pmol, 1 pmol/ml) for 2 h at room temperature in 6 x SSC, 10 x Denhardt's, 50 μg/ml single-stranded DNA. The filters were washed first at 50°C in 6 x SSC, 0.1% SDS and then (**a**) for 2 min at 60°C or (**b**) for 2 min at 65°C. The filters were then exposed to pre-flashed X-ray film for 16 h at −70°C.

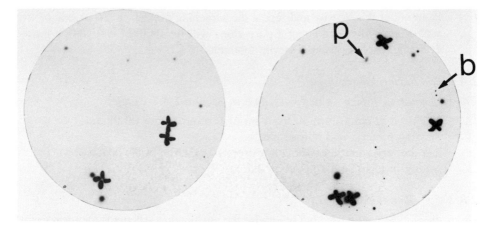

Figure 3. Use of duplicate filters in screening a phage λ library. A genomic library of *Xenopus laevis* DNA in the phage λ vector Charon 4A (16) was screened with a nick-translated *X. laevis* β-globin cDNA clone. Areas of the original master plates containing positive plaques were picked, titrated, and plated out at a density of about 100 plaques per plate. Two duplicate filters from each plate were hybridised with the nick-translated DNA probe (5 x 10^7 c.p.m/μg), washed in 1 x SET buffer at 68°C and exposed to pre-flashed X-ray film at −70°C overnight with an intensifying screen (see *Table 3*). Six plaques are positive on both filters. Note the characteristic comet shapes of some of these authentic positives (**p**) compared with other 'background' spots (**b**).

with the filters bearing the DNA or, alternatively, by filtering the probe through nitrocellulose.

2.4 Purification of Positive Clones

After the X-ray films have been examined and authentic positive signals identified, the bacterial colonies or phage plaques giving positive signals must be purified. If the original bank was plated at low density, the positive plaque or colony may be identified unambiguously and isolated free of neighbouring clones. However, even when this is possible it is still a good idea to plate it out again at low density (50−100 per 9 cm plate) and repeat the binding and hybridisation schedule to check that all the colonies give a positive signal. In contrast, when large genomic or cDNA libraries are being screened, the high cell or plaque densities that must be used (Sections 2.1.2 and 2.1.3) preclude the possibility of picking a single colony or plaque in pure form. In these cases the following procedures are used to purify the positives.

2.4.1 *For Bacterial Colonies*

(i)   Align the autoradiograph with the master plate using the asymmetric pattern of ink marks keyed in the plate [Section 2.1.2, step (iv)] and the signals from the radioactive ink on the film.

(ii)   Identify an area of the plate that has given rise to a positive signal and recover a plug of agar using the wide end of a sterile glass Pasteur pipette. Shake the plug into 1 ml of L broth [see Section 2.1.2, step (i) for composition] in a sterile tube and vortex briefly.

(iii) Plate serial dilutions on L agar plates to determine the dilution required to give 50−200 colonies per 9 cm diameter plate.

(iv) Plate out at this density and repeat the screening procedure. After one or two such rounds of re-screening a pure clone will be obtained and can be picked as a single colony using a sterile toothpick.

2.4.2 *For Bacteriophage* λ

The procedure is similar to that described in Section 2.4.1 except:

(i) add 10 μl of chloroform to the 1 ml of L broth before titration to lyse the remaining bacteria and release phage;

(ii) mix the serial dilutions with a few drops of a culture of the host bacteria before plating in top agar (see Section 2.1.3).

2.4.3 *For Bacteriophage M13*

Chloroform must not be used but otherwise the purification scheme is identical to that used for bacteriophage λ (Section 2.4.2).

3. SECONDARY SCREENING AND DNA ANALYSIS TECHNIQUES

3.1 **DNA Dot Blots**

3.1.1 *General Comments*

In this procedure DNA is immobilised on an inert support, normally a nitrocellulose filter, by direct application. The advantage of the technique is that a known amount of DNA is applied to the filter and therefore accurate quantitation is possible. Also, because many samples can be immobilised very quickly, the technique can be used to screen large numbers of samples. The major disadvantage, compared with the Southern transfer procedure (Section 3.3), is that total DNA is analysed rather than DNA fragments generated by restriction endonuclease digestion. Hence there is no simple way of distinguishing between cross-hybridisation of the probe to other nucleic acids present in the sample and hybridisation backgrounds will be higher than in Southern transfer.

3.1.2 *Sample Application Using a Manifold*

The following procedure has been developed for binding plasmid DNA to a nitrocellulose support using a multiple filtration device (e.g., the minifold apparatus of Schleicher and Schüll, Cat. SRC-96). The procedure works well with relatively impure preparations of plasmid DNA such as those obtained by rapid, small-scale isolation procedures (e.g., ref. 18). In the first step the sample is treated at high temperature which partially depurinates the DNA (19). The DNA is then treated with alkali to nick supercoils and denature the DNA strands so that these then bind to the nitrocellulose (nicking is necessary to prevent the very rapid re-annealing that occurs when an intact supercoil is denatured and then returned to conditions which permit renaturation).

(i) Take a sheet of nitrocellulose paper and cut a segment large enough for the number of samples to be bound. Cut a piece of thick chromatography paper (e.g., Whatman 3MM paper) to the size of the manifold.

(ii) Wet the paper in 2 x SSC and lay it on the base of the manifold. Wet the nitrocellulose by dropping it onto 2 x SSC in a dish and then place it on the top of the chromatography paper.

(iii)   Cover any manifold holes that will not be required for samples with Parafilm.

(iv)   Place the top section of the manifold in position. Apply low suction from a water pump. Place 500 μl of 2 x SSC into one of the wells and determine how quickly the liquid disappears. Adjust the suction so that at least 5 min is required to aspirate the buffer through the filter. Leave the water running at this flow-rate until the samples have been bound.

(v)   Place each DNA sample into a plastic microcentrifuge tube in a final volume of 50 μl in 1 mM EDTA, 20 mM Tris-HCl, pH 7.6 and heat in a boiling water bath for 10 min. Pierce the lids of the tubes before boiling to prevent them popping open. The amount of DNA used will depend on the purpose of the experiment but the upper limit for binding to nitrocellulose is about 1 μg DNA/mm^2. Hence a dot of 4 mm diameter will bind up to about 10 μg of DNA.

(vi)   Add 50 μl of 1 M NaOH and incubate at room temperature for 20 min.

(vii)   Place the samples in an ice bucket and add 400 μl of neutralisation buffer [Section 2.1.2, step (x)]. Check one of the samples for neutrality using pH paper. *Caution: The samples should be bound to the filter as quickly as possible after neutralisation to prevent renaturation of the DNA.*

(viii)   Pipette the samples into the wells of the manifold. If the water flow-rate has been set correctly, the samples should take at least 5 min to pass through the filter [see (iv) above]. This low rate of application is required to ensure efficient binding of the DNA to the filter.

(ix)   Dismantle the apparatus. Place the filter onto a sheet of chromatography paper, allow to air-dry for 1 h and then label the nitrocellulose using a waterproof marker pen.

(x)   Bake the filter at 80°C for 2 h. The filter is now ready for hybridisation (see *Table 3* or *4*).

3.1.3 *Manual Sample Application*

If a manifold is not available, the following procedure may be used. It is more time-consuming for large numbers of samples and gives larger, more irregularly shaped 'dots'.

(i)   Wet a sheet of nitrocellulose by dropping it into 10 x SSC. Place it on a 'concertina' prepared by folding a double thickness of aluminium foil into parallel ridges separated by about 1 cm.

(ii)   Air-dry the filter *in situ* for about 1 h or under a heat lamp for a shorter time.

(iii)   Nick and denature the DNA sample as described above [Section 3.1.2, steps (v) and (vi)] but at the end of the alkali treatment neutralise by the sequential additon of 50 μl of 1 M Tris-HCl (pH 8.0), 50 μl of 1 N HCl and 10 μg of carrier tRNA.

(iv)   Recover the DNA by ethanol precipitation (*Table 5*, steps 7−9).

(v)   Dissolve the DNA in 5 μl of water and place this on ice.

(vi)   Spot 1 μl aliquots of each sample onto a marked area of the nitrocellulose filter not in contact with the aluminium foil.

(vii)   Wait until the sample has dried and then repeat the application until all the sample has been utilised.

(viii) Air-dry and bake the filter as in Section 3.1.2 [steps (ix) and (x)] and then proceed with the hybridisation (*Table 3* or *4*).

A similar protocol is also described in Chapter 4, Section 6.2.1.

3.2 Hybrid Selection

3.2.1 *General Comments*

This procedure is used to identify the polypeptide encoded by a cloned DNA fragment.
The DNA is bound to an inert support and hybridised to RNA under conditions of DNA
excess. The mRNA which hybridises to the filter is then recovered and translated in
a suitable *in vitro* protein translation system. The inert support may be either nitrocellulose (20) or chemically-activated paper (21). Nitrocellulose has a very high capacity
for DNA and it is very simple to bind DNA to the filter. However DNA is gradually
released from nitrocellulose and the filters are very fragile. Hence filters can normally be
used only two or three times. Chemically-activated paper may be re-used indefinitely
but the DNA binding capacity is lower than that of nitrocellulose and the binding procedure is more complicated.

When designing hybrid selection experiments it is important to realise that only a
small fraction of the immobilised DNA will hybridise to the mRNA. This occurs for
two reasons. Firstly, for reasons which are unclear, only a small fraction of any DNA
is available for hybridisation when immobilised on either nitrocellulose or activated
paper. Secondly, when the DNA bound to the filter is a bacterial plasmid containing
a segment of eukaryotic DNA, as it most usually will be, the inserted DNA sequences
complementary to the mRNA will constitute only a small portion of the DNA bound.
Thus for a 5-kb recombinant containing a 1-kb cDNA insert, only 10% of the input
DNA will hybridise to the complementary mRNA. Because of these two factors, and
because hybridisation rates are much lower for immobilised nucleic acids than for nucleic
acids in solution, it is necessary to use a relatively large amount (10 μg) of plasmid
DNA in the hybridisation in order to purify sufficient mRNA to give a detectable translation product. The amount of RNA required will of course vary depending upon the
abundance of the complementary mRNA but the amount we suggest (250 μg of total
cellular or cytoplasmic RNA) yields a readily detectable amount of a low to medium
abundance class mRNA (i.e., an mRNA constituting 0.1 − 1% of the population).

3.2.2 *Hybrid Selection Using Nitrocellulose*

A suitable procedure is described in *Table 5*. Plasmid DNA is immobilised onto nitrocellulose using one of the two procedures described above for preparing DNA dot blots
(Sections 3.1.2 and 3.1.3). The plasmid DNA may be prepared by a rapid procedure
such as the boiling procedure of Holmes and Quigley (18). When screening large
numbers of bacterial colonies for a particular recombinant it is possible to reduce the
scale of the operation by preparing DNA from pools of as many as 10 different recombinants and binding these to a single large nitrocellulose filter (19).

Aside from the large amount of DNA which must be used in hybrid selection, the
only modification from the dot blot procedure (Section 3.1) is to wet the filter with
sterile water after baking and then excise the DNA dots using a sterile razor blade.
When using a manifold the impression left by the rubber seal separating the

Table 5. Hybrid Selection Using Nitrocellulose Filters.

Immobilisation

1.   Wet a piece of nitrocellulose in 2 x SSCᵃ and place in position on a manifold suction device.
2.   Boil 10 μg of the cloned DNA in 50 μl of 1 mM EDTA, 20 mM Tris (pH 7.6) for 10 min.
3.   Add 50 μl of 1 M NaOH and incubate at 20°C for 20 min.
4.   Place on ice and add 400 μl of neutralisation buffer [1.4 M NaCl, 0.15 M sodium citrate, 0.25 M Tris-HCl (pH 8.0), 0.25 M HCl].
5.   Pipette the DNA sample into the manifold well under gentle suction (>5 min total passage time).
6.   Air-dry the filter for 10 min and then bake at 80°C for 2 h.
7.   Wet the filter by placing on 3MM paper saturated with water and cut the DNA dots out of the wet filter (the dry nitrocellulose is too brittle to cut cleanly).

To remove loosely-bound DNA before first use.

1.   Place each filter in 1 ml of water and incubate for 1 min in boiling water.
2.   Chill on ice and then remove the water by aspiration.
3.   Add 1 ml of water, vortex and then remove the water.
4.   Blot the filter dry on 3MM paper.

Hybridisation

1.   Place up to 20 filters in a disposable plastic scintillation vial insert with 100 μl of hybridisation bufferᵇ (50% deionised formamideᶜ, 0.9 M NaCl, 0.2% SDS, 1 mM EDTA 20 mM Pipes, pH 6.4).
2.   Incubate at room temperature for 10 min and then remove the buffer by aspiration.
3.   Add 100 μl of the hybridisation buffer containing 250 μg of the total RNA of interest.
4.   Incubate for 6 h at 37°C.
5.   Wash with 100 ml washing bufferᵈ (50% deionised formamideᶜ, 20 mM NaCl, 8 mM sodium citrate, 1 mM EDTA, 0.5% SDS) for 15 min at 37°C.
6.   Change the washing buffer and repeat step 5 for a total of five washes.
7.   Blot the filters dry on 3MM paper.

Elution of RNA

1.   Place each filter in a separate microcentrifuge tube. Add 1 ml water and vortex for 5 sec. Remove the water and repeat the wash.
2.   Add 300 μl of 1 mM EDTA to the filter.
3.   Place in a boiling water bath for 1 min.
4.   Snap freeze by plunging into dry ice.
5.   Thaw on ice and remove the filter.
6.   Add 10 μl of 1 mg/ml calf liver tRNAᵉ and sodium acetate to 0.2 M. Extract with phenol:chloroform:iso-amyl alcohol (1:1:0.04 by vol.).
7.   Add 2.5 volumes of ethanol to the supernatant and place in an ethanol-dry ice bath for 10 min. Centrifuge at 12 000 *g* for 10 min to recover the RNA.
8.   Aspirate off the supernatant. Wash the pellet twice with 95% ethanol.
9.   Dry the RNA pellet under vacuum and re-dissolve it in 5 μl water ready for translation.

Removing residual RNA from filters before re-use.

1.   Place each filter in a microcentrifuge tube and add 1 ml of 2 x SSC, 0.1 M NaOH. Leave at room temperature for 30 min and then remove the liquid by aspiration.
2.   Add 1 ml of 2 x SSC to the filter, vortex for 10 sec and then aspirate off the liquid. Repeat this wash four more times.
3.   Wash once only as in step 2 using 1 ml water and then air-dry the filters.
4.   Store the filters under vacuum at room temperature.

ᵃThe composition of SSC is given in *Table 4,* footnote (a).

ᵇHybridisation buffer can be made prior to use and stored at −20°C.

ᶜDeionised formamide is prepared as follows. Stir formamide (Analar grade) with mixed-bed ion-exchange resin [e.g. 5 g Dowex AG501, (BioRad) to 100 ml formamide] for 1−2 h at room temperature. Filter and store at −20°C.

ᵈWashing buffer must be prepared fresh using sterile solutions, glass and plastic ware.

ᵉCommercial calf liver tRNA should be purified before use as follows: phenol extract and ethanol precipitate calf-liver tRNA (Boehringer-Mannheim), re-dissolve at 1 mg/ml in water and store at −20°C.

compartments can be used to guide cutting. When using the manual sample application technique (Section 3.1.3) draw a circle 5 mm in diameter with a soft pencil before applying the DNA and ensure that all the sample is applied within this area. Once the filter circles have been cut out and before their first use, the filters should be washed to remove loosely-bound DNA (*Table 5*).

Hybridisation of the filter-bound DNA with RNA is performed in the presence of 50% formamide at 37°C and for no longer than 6 h. The low temperature and short hybridisation time are designed to minimise mRNA breakdown. Up to 20 segments of nitrocellulose (4 mm diameter) may be hybridised together in 100 µl of hybridisation buffer. Disposable plastic scintillation vial inserts (1 cm internal diameter) make ideal containers. If the RNA to be used is dissolved in water, then lyophilise the

Table 6. Hybrid Selection Using Chemically-Activated Paper.

Immobilisation

1. Sonicate 10 µg of the cloned DNA to an average size of 500 − 1000 bp or linearise the recombinant plasmid DNA with a suitable restriction enzyme.
2. Phenol extract the DNA and then recover it by ethanol precipitation. Wash the pellet with 95% ethanol and dry under vacuum.
3. Dissolve the DNA in 20 µl of 25 mM potassium phosphate buffer (pH 6.2)[a].
4. Add 80 µl DMSO, incubate at 80°C for 10 min and then cool on ice.
5. Cut circles (0.8 cm diameter) from a sheet of activated APT paper and mark them with a soft lead pencil.
6. Place a circle in a plastic scintillation vial insert and add 100 µl of the DNA solution. Incubate overnight at room temperature in the dark.
7. Remove the liquid by aspiration and wash each filter twice with 2 ml of water.
8. Add 100 µl of 0.4 M NaOH and shake at 37°C for 30 min.
9. Wash three times each with 5 ml of water.
10. Add to each filter 1 ml of 50% deionised formamide[b] buffered with 10 mM Tris-HCl pH 8.0. Store the washed filters at 4°C.

Hybridisation

The procedure is as for nitrocellulose filters (*Table 5*) except that only five filters per 100 µl may be used.

Elution

1. To each filter in a sterile plastic tube add 100 µl of elution buffer (90% deionised formamide[b], 1 mM EDTA, 0.5% SDS, 20 mM Pipes, pH 6.4). Incubate at 37°C for 30 min.
2. Transfer the eluate to a plastic microcentrifuge tube. Add to the eluate 300 µl water, 40 µl 3 M sodium acetate, pH 5.5, 10 µl calf liver tRNA[c] (1 mg/ml).
3. Add 2 volumes of ethanol and place in an ethanol-dry ice bath 10 min. Recover the RNA by centrifugation (12 000 g for 10 min).
4. Re-dissolve the RNA pellet in 250 µl 0.3 M sodium acetate, pH 5.5, and repeat the ethanol precipitation (step 3).
5. Rinse the RNA pellet twice with 70% ethanol. Dry under vacuum and dissolve the RNA in 5 µl water ready for translation.

Removing residual RNA from filters before re-use

Steps 8 − 10 of the protocol for immobilisation given above are used to regenerate the filters after each use.

[a]25 mM potassium phosphate buffer (pH 6.2): Add 3.9 ml 0.5 M KH_2PO_4 to 1.1 ml 0.5 M K_2HPO_4. Make up to 100 ml with water.
[b]To prepare deionised formamide, see *Table 5*, footnote (c).
[c]To prepare calf-liver tRNA, see *Table 5*, footnote (e).

required amount of RNA and re-dissolve it directly in hybridisation buffer. If it is stored in buffer then recover the RNA by ethanol precipitation and dissolve it in the hybridisation buffer. Hybridisation reactions should be performed in an oven or in a water bath with the tubes completely submerged (the close fit of the cap on a plastic-scintillation vial insert will prevent leakage). At the end of the hybridisation, all the filters should be put into a 500 ml beaker containing washing buffer and gently agitated at 37°C for 15 min in a shaking water bath or oven. This step is repeated five times. Then the filters are placed in separate tubes, the mRNA is eluted by denaturing the RNA-DNA hybrid and is recovered by ethanol precipitation (*Table 5*).

3.2.3 *Hybrid-Selection using Chemically-activated Paper*

Paper provides a very convenient support if the cellulose is derivatised with a chemically reactive moiety which can then be coupled to DNA or RNA. Initially a diazobenzyloxymethyl (DBM) derivative was utilised (22) but now a more convenient and stable O-aminophenylthioether (APT) derivative is more commonly used (23). Procedures for preparing and chemically activating these papers are fully described in references 22 and 23. The use of APT-paper in hybrid selection is given in *Table 6*. Its capacity is about 20 μg DNA/cm^2. Both RNA and DNA will compete for binding sites with equal efficiency and it is therefore very important that highly purified plasmid DNA is used. Up to five segments of APT-activated Whatman 540 paper (4 mm diameter) may be hybridised per 100 μl of buffer. The hybridisation conditions are identical to those used for nitrocellulose (*Table 5*) but the elution is performed in buffer containing a high concentration of formamide.

3.2.4 *Interpretation of Results and Problem Solving*

The effects of changing some of the parameters described earlier in this section are illustrated by the experiment shown in *Figure 4*. Total cellular RNA isolated from *Dictyostelium discoideum* at 3.5 h of development was hybridised to DBM-paper discs (21) bearing plasmid DNA containing a cDNA insert derived from the mRNA encoding discoidin 1a, a lectin of unknown function synthesised during *Dictyostelium* development. As mentioned in Section 3.2.1, hybrid selection should be performed under conditions of DNA excess. Lanes 6–9 show the effects of increasing amounts of total cellular RNA (12.5, 25, 50 and 125 μg, respectively, per incubation). Even at the highest level used, saturation has not been achieved. Lanes 11–13 show this in another way by increasing the amount of DNA bound to the filter with a constant amount (approximately 40 ng) of complementary RNA added in each incubation. Clearly a filter bearing 10 μg of DNA (lane 11) is as efficient in binding as filters bearing 20 μg (lane 12) or 50 μg (lane 13) of DNA. These results are consistent with the failure to obtain saturation in the RNA titration; that is, at all the ratios of RNA to DNA used, the DNA was in excess over the RNA.

The amount of RNA used in the hybridisation should be reduced or increased depending upon the expected abundance of the complementary mRNA. Up to 2 mg of RNA may be dissolved per 100 μl of hybridisation buffer and this will allow detection of lower abundance RNA sequences. As the relative abundance of the complementary mRNA decreases, the 'background' of non-specific translation products increases. Pro-

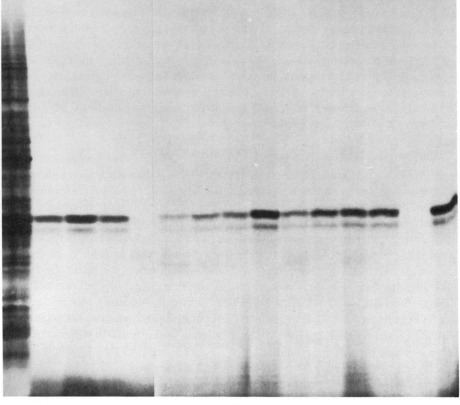

Figure 4. Hybrid-selection of a medium abundance mRNA. Total cellular RNA isolated from *D. discoideum* at 3.5 h of development was hybridised to DBM-paper discs (22) bearing plasmid DNA containing a cDNA insert for discoidin 1a (21). Hybridisations were for 4 h. Filters were washed and RNA was eluted and translated in a micrococcal nuclease-treated reticulocyte lysate (24) containing [³H]leucine. The translation products were analysed on 11−14% gradient polyacrylamide gels containing SDS and detected using fluorography (25). **Lane 1** contains the translation products directed by total cellular RNA isolated from cells at 3.5 h of development. **Lanes 2,3** and **4** contain the products directed by RNA complementary to the *Eco*RI-*Hind*III (5 kb), *Hind*III-*Eco*RI (0.33 kb) and *Hind*III-*Hind*III (0.5 kb) fragments of the plasmid (23). **Lanes 5** and **14** contain the products directed by RNA from a filter bearing heterologous DNA. Descriptions of the translation products shown in other lanes are given in the text.

vided, however, that adequate controls are performed this is not a major problem. If a cDNA clone is being analysed, the best control is to perform a parallel hybridisation with another cDNA clone in the same vector but with a different mRNA insert. The specific translation product directed by the test plasmid can then be determined by comparison with the translation products obtained with the control. This is shown by a comparison of lanes 2−4 with lanes 5 and 14 in *Figure 4*.

In the experiment shown in *Figure 4*, the amount of translatable RNA bound increased up to 6 h. By 20 h, however, some loss of translatable RNA was observed, presumably because of degradation (results not shown). Another hybridisation condition which may require alteration is the temperature. If the region of complementarity is very short or if it contains a very high proportion of A-U base pairs then the temperature of hybridisation and washing should be reduced.

Figure 4 also shows that filters which have been re-used several times are still as efficient in hybrid selection as new filters. Compare lane 10 (fresh filter bearing 10 μg DNA) with lane 11 (filter also bearing 10 μg DNA but which had been previously used in at least 25 separate hybridisation reactions).

Finally, hybrid selection can also be performed in a preparative manner. The translation product of a portion of a large (0.5 μg) batch of mRNA obtained in a preparative-scale hybridisation is shown in *Figure 4*, lane 15.

There are two common causes of failure with this procedure; RNA breakdown and inadequate sensitivity. It is obviously imperative that the RNA remains intact during the hybridisation, washing and recovery procedures. Thus the standard precautions for working with RNA (i.e., the wearing of disposable plastic or latex gloves at all times and treatment of all solutions and surfaces which contact the RNA with diethylpyrocarbonate; see ref. 26) must always be followed. RNA is most likely to be degraded during hybridisation. The simplest test for this degradation is to recover unhybridised RNA from the buffer (ethanol-precipitate the RNA after diluting 4-fold with water) at the end of the hybridisation and to compare the mean size of translation products with those obtained using unincubated RNA. The most likely cause of RNA breakdown during hybridisation is use of inadequately de-ionised formamide (see *Table 5* footnote c for a suitable method). The longer the mRNA, the more sensitive it is to single nucleolytic events which will render it biologically inactive. In these cases it may be necessary to decrease the hybridisation time considerably (perhaps as short as 1 − 2 h) even though this will entail a loss in sensitivity. Thus Smith *et al.* (27) reduced the hybridisation time to 3 h when using a 6.4 kb *Xenopus* vitellogenin mRNA.

As mentioned above, the other major cause of failure is an inadequate level of sensitivity. Provided the binding of DNA to the filter is efficient and sufficient RNA has been used, then this will normally be the result of poor translation. The most sensitive *in vitro* translation system is the nuclease-treated reticulocyte lysate. We have found the lysates sold by Amersham International and New England Nuclear to give an adequate level of sensitivity.

3.3 Southern Transfer

3.3.1. *Basic Procedure*

Southern blotting (Southern transfer) is a widely used analytical tool originally devised by Southern (28). Individual fragments of DNA, usually produced by restriction enzyme digestion, are separated by gel electrophoresis and then transferred to an inert support such as nitrocellulose. The immobilised DNA fragments can then be analysed by hybridisation with a suitable radioactively-labelled probe. It is beyond the scope of this chapter to describe agarose gel electrophoresis and so the reader is referred to a guide to this technique by Sealey and Southern (29). The detailed protocol for Southern transfer is described below.

(i)   After staining and photographing the gel (see ref. 28 for procedures), transfer it to a plastic box and immerse it in about 5 − 10 times its volume of 0.25 M HCl.

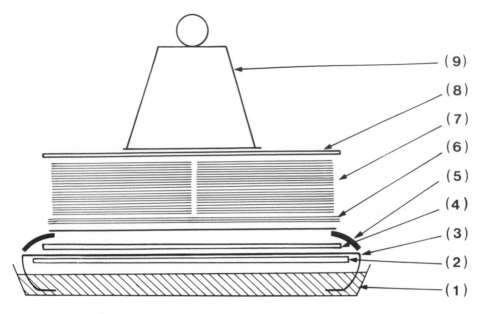

Figure 5. Cross-section of a Southern transfer apparatus. (**1**) tray filled with 20 x SSC, (**2**) glass plate (supported by two sides of the tray), (**3**) wick of three sheets of Whatman 3MM paper, (**4**) gel, (**5**) Parafilm round all sides of gel, (**6**) three sheets of Whatman 3MM paper, (**7**) paper towels, (**8**) glass plate, (**9**) weight.

Incubate for 30 min at room temperature with gentle shaking. This treatment with HCl partially depurinates the DNA, breaking it down into smaller pieces that are transferred more quickly. This step is therefore especially important for quantitative transfer of high molecular weight DNA to be achieved. However, the step can be omitted if the experiment does not require efficient transfer of DNA fragments larger than about 5 kb.

(ii)   Carefully pour off the HCl solution. Add 5−10 gel volumes of denaturing solution (0.6 M NaCl, 0.2 M NaOH) and incubate for 30 min at room temperature with gentle shaking.

(iii)   Replace the denaturing solution with the same volume of neutralising solution (1.5 M NaCl, 0.5 M Tris-HCl, pH 7.0) and shake gently at room temperature for a further 30 min.

(iv)   Set up the apparatus as shown in *Figure 5* using a glass or plastic container. This consists of three sheets of Whatman 3MM paper saturated with 20 x SSC lying on a clean horizontal glass plate. The 3MM paper dips into a reservoir containing 20 x SSC.

(v)   Place the gel, after neutralisation [step (iii)] onto the saturated 3MM paper and smooth it down with a gloved hand to remove any air bubbles trapped between gel and paper. Place Parafilm or cling-film along each edge of the gel, flush with the edge to ensure that the nitrocellulose filter will not be in direct contact with the 3MM paper.

(vi)   Cut a piece of nitrocellulose (0.45 μm pore size) about 3 cm longer and wider than the gel. Float it carefully onto the surface of some 4 x SSC in a plastic or glass box. The filter should wet evenly. When fully wetted, shake the box to

134

immerse the filter. Place the wetted filter on top of the gel and smooth it down to ensure that no air bubbles are trapped between the filter and the gel.

(vii) Place three sheets of Whatman 3MM paper cut to the same size as the filter on top of the filter. Stack some dry paper towels on top of the 3MM paper. Place a glass plate on top of the towels and a weight (0.5 − 1 kg) on top of the plate to compress the whole sandwich. Allow the DNA fragments to transfer over-night. Buffer is drawn from the reservoir through the gel and then through the nitrocellulose filter into the paper towels. As it flows, it carries with it the DNA fragments which become bound to the nitrocellulose filter.

(viii) Remove the paper towels and carefully take off the 3MM paper. At this stage the impression of the gel wells can be clearly seen on the nitrocellulose filter. Mark these with a soft pencil or water-resistant marker. To orientate the filter, cut off one corner. Carefully peel off the filter and transfer it into a dish of 4 x SSC and leave for a few minutes at room temperature. The gel should be re-stained with ethidium bromide after Southern transfer to verify that no DNA remains untransferred.

(ix) Dry the filter by first blotting with towels and then leaving at room temperature or 37°C for 15−30 min. Sandwich the filter between 3MM paper and bake at 80°C for 2−4 h [Section 2.1.2, step (xii)]. The filter is then ready for hybridisation (see Sections 3.3.2−3.3.4).

When analysing *cloned DNA*, the amount applied to a gel (usually 0.1−1 μg) is great-ly in excess of what is required to give a strong signal after hybridisation with a label-led probe. Hence, when the DNA is to be screened with several different probes, several replicate filters can be prepared from one gel using Southern blotting by laying fresh filters onto the gel in sequence. This can only be done efficiently after HCl treatment. In order to get three filters from one gel, leave the filters on the gel for 30 min, 1 h and then overnight, respectively. In contrast, when *total genomic DNA* is being analys-ed, the amount of DNA on the gel is limiting and about 10 μg of mammalian DNA must be transferred to a single filter to enable detection of a single copy sequence (pro-portionately less DNA can be used for organisms with smaller genome sizes).

3.3.2 *Hybridisation Using Dextran Sulphate*

The inclusion of dextran sulphate in the hybridisation increases the rate of hybridisa-tion about 10-fold and can increase the sensitivity about 100-fold (30). The latter is probably due to the formation of networks of probe DNA. Using this procedure, and a probe nick-translated to a specific activity of greater than 5 x 10^7 c.p.m./μg, a single copy sequence in DNA from a higher eukaryote can readily be detected after a 1 day exposure to X-ray film. Backgrounds tend to be higher when using dextran sulphate and so it is not recommended for screening libraries. The authors have used the following protocol devised by Jeffreys *et al.* (31) with satisfactory results in whole genome blot-ting experiments. The individual manipulations are described in detail in Section 2.2.

(i) Wet the filter in 3 x SSC.

(ii) Incubate the filter for 1 h at 65°C with 10 x Denhardt's solution (*Table 3*) in 3 x SSC.

(iii)   Pre-hybridise the filter for 1 h at 65°C with 10 x Denhardt's solution, 0.1% SDS, 50 μg/ml denatured salmon sperm DNA (*Table 3*).

(iv)   Incubate for 1 h at 65°C in the same solution containing 9% dextran sulphate (500 000 daltons; from Sigma). This step is best carried out in the plastic bag to be used for hybridisation. The dextran sulphate is dissolved in water as a 50% solution and kept at −20°C to prevent bacterial growth.

(v)   Denature the probe by heating at 100°C for 3 min in a small volume (0.1 ml) of water in a microcentrifuge tube. Cool the sample rapidly on ice and add 1 ml of the dextran-containing hybridisation solution. Place the mixture in the bag, remove bubbles and re-seal the bag. Mix the probe and hybridisation buffer by squeezing and inverting the bag. Incubate at 65°C overnight (see *Table 3* for details).

(vi)   The washing conditions after hybridisation and autoradiography are as described in *Table 3*, steps 10−15.

3.3.3 *Hybridisation Using Low Complexity Probes*

When the filter-bound DNA and the probe are both of low complexity (e.g., when cloned DNA is used to probe cloned DNA) follow the pre-hybridisation, hybridisation and washing procedures given in *Table 3*, but note the following points.

(i)   The specific activity of the probe needs to be only 10^7 c.p.m./μg to give a strong signal after overnight exposure. If a higher specific activity probe is used, exposure times can be reduced.

(ii)   If required, the hybridisation time can be reduced to a few hours ($\sim$ 2 x $C_0t_{1/2}$ of the probe).

(iii)   If the probe is not expected to form a perfect hybrid with the filter-bound DNA, reduce the stringency of washing and hybridisation as described in Section 2.2.1.

(iv)   If the probe contains segments of poly(dA)/(dT) or poly(dG)/(dC) homopolymer, add 10 μg/ml poly(A) or 10 μg/ml poly(C) to the pre-hybridisation and hybridisation buffers to prevent spurious hybridisation (see Section 2.3.2).

3.3.4 *Hybridisation Using High Complexity Probes*

When cloned DNA is bound to the filter, rare transcripts can be assigned to regions of the DNA by using labelled mRNA or cDNA made against the total mRNA population as the probe. In these experiments:

(i)   The concentration of radioactive probe in the hybridisation solution should be as high as possible. For example, when the probe is present at 10^7 c.p.m./ml, sequences constituting only 0.01% of the mRNA population can be detected.

(ii)   When using cDNA or mRNA probes, poly(A) (10 μg/ml) should be added to the pre-hybridisation and hybridisation buffers to prevent spurious hybridisation (see Section 2.3.2.).

(iii)   The hybridisation time should be increased to 24−48 h.

4. REFERENCES

1.   Grunstein,M. and Hogness,D. (1975) *Proc. Natl. Acad. Sci. USA,* **72,** 3961.
2.   Hanahan,D. and Meselson,M. (1980) *Gene,* **10,** 63.

3. Twigg,A.J. and Sherratt,D. (1980) *Nature, **283**,* 216.
4. Vieira,J. and Messing,J. (1982) *Gene,* **19**, 259.
5. Benton,W.D. and Davis,R.W. (1977) *Science (Wash.),* **196**, 180.
6. Messing,J. and Vieira,J. (1982) *Gene,* **19**, 269.
7. Mangiarotti,G., Chung,S., Zuker,C. and Lodish,H. (1981) *Nucleic Acids Res.,* **9**, 947.
8. Denhardt,D.T. (1966) *Biochem. Biophys. Res. Commun.,* **23**, 641.
9. Laskey,R.A. and Mills,A.D. (1977) *FEBS Lett.,* **82**, 314.
10. Wallace,R.B., Schaffer,J., Murphy,R.F., Bonner,J., Hirose,T. and Itakura,K. (1979) *Nucleic Acids Res.,* **6**, 3543.
11. Wallace,R.B., Johnson,N.J., Hirose,T., Miyake,M., Kawashima,E.H. and Itakura,K. (1981a) *Nucleic Acids Res.,* **9**, 879.
12. Woods,D.E., Markham,A.E., Richer,A.T., Goldberger,G. and Colten,H.R. (1982) *Proc. Natl. Acad. Sci. USA,* **79**, 5661.
13. Suggs,S.V., Hirose,T., Miyake,T., Kawashima,E.H., Johnson,M.J., Itakura,K. and Wallace,R.B. (1981) in *Developmental Biology Using Purified Genes,* Brown,D.D. and Fox,C.F. (eds.), p.683.
14. Wallace,R.B., Schold,M., Johnson,M.J., Dembek,P. and Itakura,K. (1981b) *Nucleic Acids Res.,* **9**, 3647.
15. Zoller,M.J. and Smith,M. (1982) *Nucleic Acids Res.,* **10**, 6487.
16. Blattner,F.R., Williams,B.G., Blechl,A.E., Denniston-Thompson,K., Faber,H.E., Furlong,L.-A., Grunwald,D.J., Kiefer,D.O., Moore,D.D., Sheldon,E.L. and Smithies,O. (1977) *Science (Wash.),* **196**, 161.
17. Williams,J.G. and Lloyd,M.M. (1979) *J. Mol. Biol.,* **129**, 19.
18. Holmes,D.S. and Quigley,M. (1981) *Anal. Biochem.,* **114**, 193.
19. Parnes,J.R.B., Velan,A., Felsenfeld,L., Ramanathan,U., Ferrini,U., Apella,E. and Lidman,J.G. (1981) *Proc. Natl. Acad. Sci. USA,* **78**, 2253.
20. Ricciardi,R.P., Miller,J.S. and Roberts,B.E. (1979) *Proc. Natl. Acad. Sci. USA,* **76**, 4927.
21. Williams,J.G., Lloyd,M.M. and Devine,J.M. (1979) *Cell,* **17**, 903.
22. Alwine,J.C., Kemp,D.J., Parker,B.A., Reiser,J., Renart,J., Stark,G.R. and Wahl,G.M. (1979) in *Methods in Enzymology,* Vol. **68**, Wu,R. (ed.), Academic Press Inc., London and New York, p.220.
23. Seed,B. (1982) *Nucleic Acids Res.* **10**, 1799.
24. Pelham,H.R.B. and Jackson,R.J. (1976) *Eur. J. Biochem.,* **131**, 289.
25. Hames,B.D. (1981) in *Gel Electrophoresis of Proteins − A Practical Approach,* Hames,B.D. and Rickwood,D. (eds.), IRL Press, Oxford and Washington D.C., p.1.
26. Clemens,M.J. (1984) in *Transcription and Translation − A Practical Approach,* Hames,B.D. and Higgins,S.J. (eds.), IRL Press, Oxford and Washington D.C., p.211.
27. Smith,D.F., Searle,P.F. and Williams,J.G. (1979) *Nucleic Acids Res.,* **6**, 487.
28. Southern,E.M. (1975) *J. Mol. Biol.,* **98**, 503.
29. Sealey,P.G. and Southern,E.M. (1982) in *Gel Electrophoresis of Nucleic Acids − A Practical Approach,* Rickwood,D. and Hames,B.D. (eds.), IRL Press, Oxford and Washington D.C., p.39.
30. Wahl,G.M., Stern,M. and Stark,G.R. (1979) *Proc. Natl. Acad. Sci. USA,* **76**, 3683.

CHAPTER 6

Hybridisation in the Analysis of RNA

JEFFREY G. WILLIAMS and PHILIP J. MASON

1. INTRODUCTION

Nucleic acid hybridisation can be used to determine the primary structure, the concentration and the rate of synthesis of an individual RNA species in a complex population. Many of the procedures used are very similar to those described in Chapter 5 for DNA. However, RNA molecules are much more susceptible to degradation than DNA. The major danger is of cleavage by ribonucleases during extraction and purification. Therefore procedures to minimise this problem should always be followed (1). In addition, at high temperature, and especially at neutral or alkaline pH, RNA is degraded much more rapidly than DNA. Hence hybridisation should be performed for the minimum possible time in a buffer of slightly acidic pH at the lowest temperature allowing efficient hybridisation. Hence formamide-containing buffers are a necessity rather than an option for certain hybridisation procedures where RNA integrity is crucial.

Four procedures are commonly used to analyse RNA sequences.

(i) *Northern transfer* can be used to give an estimate of the length of an RNA transcript.

(ii) *Nuclease S1 mapping* can be used to determine the position of the 5' and 3' termini of a gene and of any introns within it.

(iii) *Primer extension* can be used to determine the position of the 5' terminus of a gene and to identify the transcripts of related genes.

(iv) *RNA dot blots* can be used to give an estimate of the relative concentration of a specific mRNA in an mRNA population.

All of these techniques can also be used quantitatively to determine the relative concentration of an RNA sequence in different RNA populations.

2. NORTHERN TRANSFER

2.1 General Comments

Northern transfer is the equivalent for RNA of the Southern transfer procedure for DNA analysis (Chapter 5, Section 3.3). The RNA is separated by electrophoresis on an agarose gel under denaturing conditions, transferred to a nitrocellulose filter and specific RNA species are detected by hybridisation with a radioactively-labelled probe. The technique is extremely sensitive. The intensity of the autoradiographic signal is a measure of the concentration of the specific RNA and the position of migration is a measure of its molecular weight. Total cellular or cytoplasmic RNA can be used.

The gel electrophoresis step is best performed using conditions which disrupt RNA secondary structure. This greatly improves resolution and allows an accurate estimation of the length of the RNA molecule. Since denatured RNA and DNA molecules of the same size migrate with similar mobilities, it is possible to use DNA restriction fragments as size markers. A number of methods of denaturation are in common use, for example, the use of glyoxal (2), formaldehyde (3) and methyl mercuric hydroxide (4). Only the glyoxal procedure will be presented here since we have found this to be as effective as the other two techniques both of which involve the use of toxic compounds.

In the first Northern transfer experiments, RNA was bound to diazotised paper (5). Nitrocellulose was not used because it was thought that RNA would not bind efficiently. However, Thomas (6) has shown that denatured RNA will bind very efficiently to nitrocellulose at high ionic strength. In fact, due to the higher binding capacity of nitrocellulose compared with diazotised paper, much greater sensitivity can be achieved, as little as 1 pg of specific RNA being readily detectable. Thus, nitrocellulose (or one of the alternative supports which are more robust) is now almost always used for Northern transfer and only this procedure will be described.

The RNA is denatured by incubation at 50°C with glyoxal and dimethylsulphoxide (DMSO). The high temperature and DMSO combine to disrupt hydrogen bonding which allows glyoxal to interact with the RNA. It modifies guanine residues to form a covalent adduct which is stable at neutral or acidic pH. After electrophoresis and transfer to nitrocellulose, the glyoxalation reaction is reversed by a high temperature incubation at pH 8.0 and the specific RNA is detected by hybridisation with a radioactively-labelled probe. Because glyoxal also interacts with proteins, it is important to remove all protein as this can cause nucleic acids to become stuck at the top of the gel. The procedure described is based on methods developed by Thomas (7).

2.2 Procedure

Glyoxal is readily oxidised to glyoxylic acid which will cause RNA degradation during the denaturation step. Hence it is crucial to remove this impurity from the glyoxal by de-ionisation (*Table 1*). It is convenient to de-ionise a large volume of glyoxal and to store it frozen in aliquots at −20°C. Excess glyoxal should not be re-frozen − use

Table 1. Deionisation of Glyoxal.

1.	Place 20 ml of 40% glyoxal solution (BDH Chemicals Ltd.) in a 100 ml beaker and add 20 g of AG501-X8 ion-exchange resin (BioRad).
2.	Stir the suspension at room temperature for 30 min or until the beads are completely decolourised.
3.	Transfer the supernatant to a 50 ml beaker containing 4 g of fresh AG501-X8 resin and stir at room temperature for 30 min. If the pH (measured using indicator paper) is between pH 5.5 and 6.0, proceed to step 4. If not, transfer the supernatant to a beaker containing another 4 g of fresh resin and repeat this step.
4.	Pour the supernatant through a Pasteur pipette containing a filter of loosely-packed glass wool.
5.	Freeze the glyoxal in 100 μl aliquots at −20°C in 0.4 ml plastic microcentrifuge tubes.

Stored at −20°C glyoxal is stable for many months. However it is necessary to check the pH of each aliquot before use. If it is lower than pH 5.5 then the aliquot should be de-ionised again by the addition of a small amount (~20 μl packed volume) of AG501-X8 resin. Incubate this at room temperature for 10 min, vortexing the mixture every 2 or 3 min. Centrifuge the tube at 12 000 g for 1 min, and remove the supernatant to a fresh tube ready for use.

Table 2. Denaturation of RNA, Gel Electrophoresis and Northern Transfer to Nitrocellulose.

1.	To the RNA (≤ 20 μg) in 5 μl of water in a plastic microcentrifuge tube add:

		Final concentration
	4 μl of de-ionised glyoxal (*Table 1*)	1 M
	3 μl of 80 mM sodium phosphate buffer (pH 6.5)	10 mM
	12 μl of DMSO (BDH Chemicals Ltd.)	50%

2.	Seal the tube and heat to 50°C for 1 h in a covered water bath.
3.	Chill on ice and add 5 μl of gel sample buffer[a] (0.05% bromophenol blue, 50% sucrose, 10 mM sodium phosphate, pH 6.5).
4.	Prepare a horizontal 1% agarose gel (low electro-endosmosis grade agarose from Bethesda Research Laboratories) for the analysis of RNA larger than 1 kb or a 1.5% agarose gel for smaller RNA. Use 10 mM sodium phosphate buffer (pH 6.5) to prepare the gel which should be 3 mm in depth with sample slots 0.8 cm wide.
5.	Once set, place the gel in the electrophoresis apparatus using 10 mM sodium phosphate buffer (pH 6.5) as electrophoresis buffer. Load the RNA samples. Also load DNA restriction fragments as size markers in another well. Preferably these should be radioactively-labelled but unlabelled markers can also be used (see Section 2.2). The DNA markers must be glyoxalated in exactly the same way as the RNA samples (steps 1−3) before use.
6.	For a gel 20 cm long, perform the electrophoresis at 100 V for 3−4 h by which time the marker dye will have migrated approximately 15 cm. Re-circulate the buffer during electrophoresis (see Section 2.2). However, do not attempt to re-circulate the buffer until the samples have electrophoresed a few millimetres into the gel or they will be displaced into the running buffer.
7.	If radioactively-labelled DNA markers have been used, wet a sheet of nitrocellulose in water, and set up the whole gel for transfer using 20 x SSC[b] exactly as described in Chapter 5, Section 3.3. Alternatively, if unlabelled marker have been used, excise that segment of the gel containing the markers and locate their positions as described in the text. Then set up the remainder of the gel, containing the separated RNA, for Northern transfer.
8.	Transfer for at least 12 h and then air-dry and finally bake the filter at 80°C for 2 h [see Chapter 5, Section 2.1.2, step (xii)]. *Note:* It is important *not* to wash the blot in low salt buffer (eg., 2 x SSC) before baking since this will cause RNA to be lost from the filter.
9.	Place the filter in 200 ml of 20 mM Tris-HCl, pH 8.0 at 100°C and allow to cool to room temperature. This removes residual glyoxal.
10.	Detect the RNA species of interest by hybridisation as described in *Table 3*.

[a]Store this buffer at −20°C.
[b]The composition of SSC is 0.15 M NaCl, 0.015 M trisodium citrate.

a fresh aliquot for each experiment.

The procedures for denaturation of RNA using de-ionised glyoxal, gel electrophoresis and Northern transfer to nitrocellulose are all described in *Table 2*

The samples are stable for many hours after denaturation and can be kept at 4°C until electrophoresis. The electrophoresis is carried out using either a 1.0 or 1.5% horizontal agarose slab gel depending on the size of the RNA. The gel is run in a phosphate buffer of low ionic strength. Because of the low buffering capacity, it is necessary to re-circulate this buffer during electrophoresis. This is very important as glyoxal will dissociate from the RNA if the pH rises above about pH 8.0. This is best done by supporting the gel tank on two magnetic stirring plates with a magnetic follower located near the anode and the cathode and re-circulating the buffer between the anodic and cathodic chamber using a peristaltic pump. In order to prevent the gel floating up, lay a glass rod down each side of the gel. Electrophoresis is continued until a marker dye (bromophenol blue) contained in the RNA sample has migrated three-quarters of the way through the gel.

After electrophoresis, the transfer of the RNA to nitrocellulose is maximally efficient when the gel contains a low salt concentration and the transfer buffer contains a very high salt concentration (7). Alkali treatment of the gel to remove glyoxal residues or staining with ethidium bromide to locate the separated RNA greatly reduce the efficiency of transfer. Hence instead it is desirable to run radioactively-labelled markers (e.g., end-labelled DNA restriction fragments; see Appendix II). These are transferred to the nitrocellulose along with the separated RNA and can be located after hybridisation and eventual autoradiography. If no suitable labelled markers are available, then unlabelled DNA restriction fragments (Appendix II) must be used as markers. The marker DNA fragments should be denatured with glyoxal in the same way as the RNA samples. Because single-stranded DNA binds much less ethidium bromide than double-stranded DNA, it is necessary to use at least five times more marker DNA than would be required for visualisation of DNA separated by gel electrophoresis using non-denaturing gels. After electrophoresis, cut off a segment of gel containing the separated markers and locate these as follows.

(i)   Place the segment of gel in greater than 10 gel volumes of 50 mM sodium hydroxide and shake gently for 30 min at room temperature to de-glyoxalate the markers.

(ii)   Neutralise by gently agitating for 10 min in more than 20 gel volumes of 50 mM sodium phosphate, pH 6.5. Repeat this twice more. Include ethidium bromide at 1 μg/ml in the last change of buffer.

(iii)   Photograph the segment under u.v. illumination (8).

(iv)   The remainder of the gel should be set up on a transfer apparatus using 20 x SSC exactly as described in Chapter 5, Section 3.3.1.

The transfer of RNA is complete within about 12 h, and the blots are then air-dried and baked at 80°C for 2 h (*Table 2*, step 8). This partly removes the glyoxal and also fixes the RNA to the nitrocellulose. After baking, the remainder of the glyoxal is removed by incubating the filter in Tris-HCl buffer, pH 8.0, at 100°C (*Table 2*, step 9). Finally, hybridisation is performed in formamide using a minor modification of the method of Wahl *et al.* (9) as described in *Table 3*.

2.3 Potential Problems

The most common problem with Northern transfer is an unacceptable level of degradation of the RNA. A low level of degradation which results in a 'shadow' migrating ahead of the intact RNA is quite common. However, in extreme cases all the RNA will migrate as a diffuse smear of low molecular weight. This is frequently caused by inadequate de-ionisation of the glyoxal but if the procedure described in *Table 1* has been followed exactly this is unlikely to be the explanation. Alternatively, it may arise from RNase contamination in one of the solutions and replacing all these may solve the problem. Note that RNase is sometimes used to destroy RNA in 'mini-preparations' of bacterial plasmids before electrophoresis (e.g., see ref. 10). We have found that a gel box used for analysing such preparations retains sufficient RNase to degrade RNA if the box is subsequently used for RNA electrophoresis. Hence it is advisable to reserve apparatus solely for RNA electrophoresis and to treat it with 0.2% diethyl pyrocarbonate before use. If none of these measures proves effective then it may be that the

Table 3. Hybridisation after Northern Transfer.

1.	Place the filter in 30−40 ml of pre-hybridisation buffer in a plastic bag and seal it (Chapter 5, *Table 3*). Pre-hybridisation buffer has the following composition:
	  50% formamide[a]
	  5 x SSC[b]
	  50 mM sodium phosphate buffer, pH 6.5
	  250 µg/ml sonicated denatured salmon sperm DNA[c]
	  10 x Denhardt's solution[d]

2.	Incubate with gentle agitation at 42°C for 8−20 h.
3.	Discard the pre-hybridisation buffer and replace with hybridisation buffer comprising four parts pre-hybridisation buffer and one part of 50% (w/v) dextran sulphate solution[e].
4.	Denature the nick-translated probe (Chapter 5, *Table 3*) and add it to the bag. The probe should be at a specific activity of $>10^8$ c.p.m./µg and at a concentration of approximately 5 ng/ml.
5.	Re-seal the bag and mix the probe thoroughly with the hybridisation buffer by gently squeezing and inverting the bag.
6.	Hybridise for approximately 20 h at 42°C with gentle agitation.
7.	Wash the filter in four changes of 2 x SSC containing 0.1% SDS for 15 min at 50°C.
8.	Wrap the filter in plastic film while still damp and expose to pre-flashed X-ray film at −70°C using an intensifying screen.
9.	To re-use the filter, remove the probe by placing the filter in a tray containing sterile distilled water at 100°C and leave it for 5−10 min or until it cools to room temperature.

[a]See Chapter 5, *Table 5* footnote (c) for preparation.
[b]See *Table 2*, footnote (b) for composition.
[c]See Chapter 5, *Table 3* footnote (b) for preparation.
[d]See Chapter 5, *Table 3* footnote (a) for composition.
[e]See Chapter 5, Section 3.3.2.

RNA was actually isolated in a degraded form. This can be easily determined by analysing an RNA sample which is *known* to be intact in parallel with the suspect RNA. Indeed, it is a good idea routinely to include a sample of intact total cellular RNA, which can be visualised by ethidium bromide staining of the corresponding segment of the gel. To do this, place the gel segment in 0.5 M ammonium acetate containing 5 µg/ml ethidium bromide. After 30−60 min, de-stain the gel by brief washing in water. View the stained RNA by u.v. transmission (8).

3. RNA DOT BLOTS

The principle of this technique is identical to that of the DNA dot blot method (Chapter 5, Section 3.1). A known amount of RNA is spotted onto an inert support, such as nitrocellulose, and the amount of specific RNA is determined by hybridisation with a suitable radioactively-labelled probe. As with DNA dot blots, the technique is rapid, very sensitive and can be made semi-quantitative if sufficient radioactivity is hybridised to allow scintillation counting of the radioactive 'dots'. However, it provides no information as to the number or size of RNA species hybridising to the probe. The procedure in *Table 4* is based on the method described by Thomas (7).

4. NUCLEASE S1 MAPPING

4.1 **General Comments**

The enzyme nuclease S1 degrades single-stranded DNA and RNA but at low temperature and in high ionic strength buffers it does not digest double-stranded nucleic acids at

Table 4. RNA Dot Blots.

1.	Place up to 10 μg of RNA in 2 μl of water in a 0.4 ml microcentrifuge tube and add 2 μl of denaturation solution[a].
2.	Denature the RNA by incubation for 1 h at 50°C. Ensure that the tube is well submerged and cover the water bath to minimise evaporation.
3.	After the incubation, centrifuge briefly to bring the contents to the bottom of the tube.
4.	Prepare suitable dilutions of the samples in 0.1% SDS, each dilution to be 4 μl final volume.
5.	Take a sheet of nitrocellulose large enough for the number of dots to be applied, float it onto water then place it in 20 x SSC[b] for 5 min at room temperature.
6.	Dry the filter on a 'concertina' of foil (Chapter 5, Section 3.1.3) under a heat lamp.
7.	Leaving the filter on the foil, apply the 4 μl of each sample in one application.
8.	Air-dry the filter then bake at 80°C for 2 h.
9.	Place the filter in 200 ml of 20 mM Tris-HCl, pH 8.0, at 100°C and allow to cool to room temperature to remove residual glyoxal.
10.	Hybridise and wash the filters as described in *Table 3*.

[a]For each 100 μl of denaturation solution mix: 34 μl of de-ionised glyoxal (*Table 1*); 20 μl of 0.1 M sodium phosphate (pH 6.5); 46 μl of water.
Note: DMSO should not be included as it will dissolve nitrocellulose.
[b]See *Table 2*, footnote (b) for composition.

any appreciable rate. In S1 mapping, RNA is hybridised to a DNA molecule which is complementary to the RNA over only part of its length. After hybridisation, the reaction mix is incubated with nuclease S1 which degrades unhybridised segments of the DNA probe to leave discrete DNA fragments. The size of these fragments is equal to the length of the nucleotide sequence over which there is perfect homology between the RNA and DNA. This method was initially used to determine the location and size of intervening sequences in eukaryotic mRNA (11). The DNA fragments were resolved by agarose gel electrophoresis and individual fragments were then identified by Southern transfer. A more sensitive and accurate procedure is to use an end-labelled restriction fragment as a probe (12). The latter technique is more technically demanding but it is now much more commonly used than the original procedure.

A restriction fragment of DNA that spans the predicted RNA terminus or splice point is purified and end-labelled with [32]P (Chapter 2). This probe is then denatured and hybridised with the RNA and the resulting RNA-DNA hybrids are treated with nuclease S1. The size of the labelled DNA strand in the protected hybrid is then determined by polyacrylamide or agarose gel electrophoresis under denaturing conditions. This length corresponds to the distance between the labelled end of the DNA restriction fragment and the RNA terminus or splice point. (*Figure 1*). Moreover, under conditions of DNA excess, the intensity of the autoradiographic signal from this labelled DNA fragment is directly proportional to the concentration of the hybridising RNA species.

We have obtained the most unambiguous, sensitive and reproducible results using relatively short (100−600 nt) single-stranded DNA probes. The alternative is to use a double-stranded probe and perform the hybridisation in 80% formamide when DNA renaturation is suppressed. The protocol for S1 mapping used by the authors for single-stranded probes is given in Section 4.2. The modifications needed for double-stranded probes using the formamide method are described in Section 4.3.

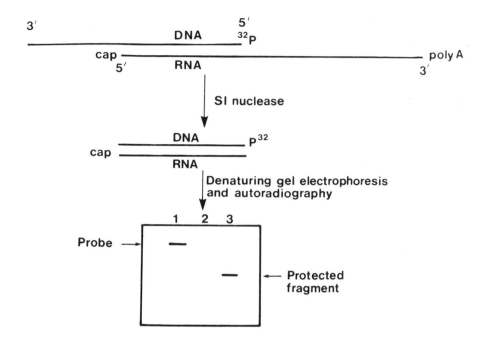

Figure 1. The principle of nuclease S1 mapping using an end-labelled probe. The diagram illustrates how the 5′ terminus of a polyadenylated mRNA might be determined using a single-stranded DNA probe. Normally the probe would be a fragment derived from a cloned copy of the gene by cleavage with appropriate restriction enzymes. The schematic representation of the autoradiogram (lower part of figure) demonstrates the results expected from a typical experiment. **Lane 1**, the probe without nuclease S1 digestion; **lane 2**, probe incubated under hybridisation conditions without complementary RNA and then digested with nuclease S1; **lane 3**, probe hybridised with complementary RNA and digested with nuclease S1 leaving the fragment protected from nuclease digestion by hybridised RNA.

4.2 Single-stranded Probes

4.2.1 *Choice of Probe*

The genomic DNA must have been characterised by restriction enzyme analysis and ideally by DNA sequencing, though the latter is not essential. The approximate location of transcribed regions should have been deduced by Northern transfer, Southern blotting or a combination of both techniques. The orientation of the gene may be deduced from comparison with a cDNA clone or from Northern blotting with strand-specific probes.

By reference to the restriction map of the gene, choose a probe which spans the RNA terminus or splice point of interest but which overlaps that point by at least 50 nt. The shorter the length of the hybridising DNA the greater is the accuracy with which the RNA can be mapped. If the experiment is designed to identify the RNA termini, several probes could by used that together span the transcribed region.

When mapping a 5′ terminus, or an intron upstream of the labelling point, a 5′-labelled DNA fragment is used. Cleaner results will be obtained if an enzyme is used which leaves a projecting terminus after cleavage since the label will then be removed from

Table 5. Isolation of a DNA Fragment by Gel Electrophoresis.

1.	Digest 20−50 μg of DNA with the appropriate restriction enzymes. Check that the digestion is complete and the appropriate fragment is well isolated from other fragments on the gel by electrophoresis of an aliquot (0.5 μg) of the restricted DNA on a native polyacrylamide gel (see ref. 8 for experimental details of gel electrophoresis).
2.	Run the rest of the restricted DNA on a 3 mm thick polyacrylamide gel in 0.5 x TBE buffer[a] loading 10 μg of DNA per lane.
3.	After electrophoresis, place the gel onto a piece of cling film on a thin-layer chromatography plate (Merck 60^{F}254, cat. no. 5554). Illuminate the gel with u.v. light (~254 nm) in a dark room. Expose the gel to u.v. for as short a time as possible since prolonged exposure will damage the DNA. The DNA fragments will appear dark on a light background.
4.	Using a clean razor blade, excise the gel slice containing the fragment.
5.	Cut a length of dialysis tubing about 5 cm longer than the gel slice and seal it at one end with a plasti mM boric the bag with 0.5 x TBE buffer and place the gel slice in the bag. Gently extrude excess buffer and air bubbles then seal the bag with another clip.
6.	Arrange the bag in a flat bed electrophoresis tank with the slice to the cathodal side of the bag and at 90° to the current flow. Add sufficient 0.5 x TBE buffer to just cover the bag. Electrophorese at 100 V for a sufficient time to electrophorese the DNA out of the gel slice (1 h is sufficient for fragments <1 kb long).
7.	Reverse the direction of current flow for 30 sec to detach any DNA which has become bound to the dialysis tubing.
8.	Remove the liquid from the bag. Filter it through a plug of polyallomer wool in a 1 ml disposable pipette tip into a siliconised microcentrifuge tube.
9.	Add 0.1 volumes of 3 M sodium acetate pH 5.2 and then add 2.5 volumes of ethanol. Mix well and place in an ethanol-dry ice bath for 10 min. Recover the DNA by centrifugation (12 000 g, 10 min).
10.	Dry the DNA under vacuum and dissolve it in 20−50 μl of TE buffer (10 mM Tris-HCl, pH 7.5, 1 mM EDTA). Store at −20°C.
11.	Determine the concentration and purity of the eluted fragment by running an aliquot (~5%) on a polyacrylamide or agarose gel with a known amount of a DNA molecular weight marker in a parallel track.

[a]0.5 x TBE buffer has the composition 45 mM Tris, 45 mM boric acid, 1.25 mM EDTA.

any renatured probe during the nuclease treatment. In contrast, when 3′-labelled probes are used, and if strand separation is not perfect, any renatured probe remains labelled after S1 treatment. This problem may be overcome by judicious choice of the probe such that the region of complementarity between the DNA and the RNA is sufficiently smaller than the total length of the probe for the protected fragment to be well separated from the probe during analytical gel electrophoresis.

4.2.2 *Isolation of the Probe Fragment*

Only in exceptional cases will the position of suitable restriction sites be such that the probe can be made by labelling a total restriction digest of a plasmid clone or subclone. In most cases, where there are multiple restriction sites, this would produce a bewildering array of labelled DNA molecules and greatly complicate the subsequent analysis. Therefore, whether using a double-stranded or single-stranded probe, it usually will be necessary to purify the restriction fragment before labelling. This can be achieved by preparative gel electrophoresis as described in *Table 5*.

4.2.3 *Labelling of the Probe*

Methods for preparing 5′- and 3′-labelled probes are described in Chapter 2. For mapping the 5' terminus of an RNA or an acceptor/intron boundary (i.e., the 3' end of an intron), the probe is labelled at the 5' terminus using calf intestine alkaline phosphatase for dephosphorylation the T4 polynucleotide kinase. For mapping the 3' end of an RNA or an intron/donor boundary (the 5' end of an intron), the probe is labelled at the 3' terminus by 'filling in' using the Klenow fragment of *Escherichia coli* DNA polymerase I. It may be possible to design a labelling strategy that will result in only the correct strand being labelled. For example, phosphate groups can be removed asymmetrically by incubation with the phosphatase after treatment with one restriction enzyme but before restriction with a second one. Alternatively, when 'filling in' at 3' ends, particular combinations of nucleotide triphosphates can be used to label some restriction enzyme sites but not others by virtue of nucleotide sequence differences at those sites. If such a strategy cannot be designed then both strands will have to be prepared and used in S1 mapping, though of course only one will hybridise to the RNA and give a result. The specific activity of the probe should be as high as possible if maximal sensitivity is to be achieved.

4.2.4 *Strand Separation*

Strand separation is achieved by denaturing the DNA and resolving the two strands by gel electrophoresis under non-denaturing conditions using a vertical polyacrylamide slab gel. In our hands, this procedure is almost always successful but it relies on there being a sufficient difference in electrophoretic mobility for the two strands to be recovered separately from the gel. The difference in electrophoretic mobility results from the different secondary structure of the two strands. Hence it is important not to overheat the gel by running at too high a current as this will unfold the DNA. *Table 6* describes a protocol which we have used to separate the strands of DNA fragments of between 130 and 600 bp in length. The two rinses of the DNA probe with 70% ethanol (*Table 6*, step 4) prior to denaturation are designed to remove all traces of salt which would otherwise favour re-annealing. The gel contains a very low proportion of bis-acrylamide since this has been found empirically to increase resolution.

4.2.5 *Hybridisation and Digestion with Nuclease S1*

During the hybridisation (*Table 7*) the probe should be in molar excess over the complementary RNA species. A simple way to check this is to perform parallel hybridisations with varying amounts of RNA or probe. If the probe is present in a sufficient excess, the signal obtained will be proportional to the RNA concentration and independent of the probe concentration. In addition, the probe should be at a concentration sufficient to drive the reaction to completion. Under the conditions given in *Table 7*, 2 fmol of probe DNA in a reaction volume of 10 μl will drive the hybridisation to completion in 6 h. We normally use between 2 and 10 fmol of probe at a specific activity of 5 x 10^6 d.p.m./pmol with 5- or 10-fold excess of DNA over complementary RNA.

4.2.6 *Interpretation of Results*

The size of any labelled bands in the gel is determined by comparison with the molecular weight markers run on the same gel. This size is a measure of the distance between the labelled end of the probe and the terminus or splice junction of the RNA. Providing

Table 6. Separation of DNA Strands by Electrophoresis[a].

1.	Prepare an appropriate strand-separation gel, 40 cm long:
	For DNA fragments of >200 bp use a 5% acrylamide, 0.1% bisacrylamide gel in
	0.5 x TBE buffer[b]
	For DNA fragments of <200 bp use a 8% acrylamide, 0.24% bisacrylamide gel in
	0.5 x TBE buffer.
2.	Pre-electrophorese the gel at 5−10 V/cm for at least 1 h before use.
3.	Meanwhile, recover the labelled probe in a siliconised tube by ethanol precipitation (see *Table 5*, step 9).
4.	Rinse the DNA pellet *twice* with 70% ethanol and dry it under vacuum.
5.	Dissolve the pellet in 10 μl of denaturation buffer:
	30% (w/v) DMSO
	1 mM EDTA
	0.05% xylene cyanol
6.	Heat the mixture at 90°C for 3 min. Then chill quickly by plunging into an ice/salt/water bath.
7.	Load the sample immediately onto the strand-separation gel that has been pre-electrophoresed (step 2).
8.	Continue electrophoresis at the same voltage (5−10 V/cm) for a time sufficient to run the double-stranded form of the DNA to the bottom of the gel. The xylene cyanol will co-electrophorese with a double-stranded DNA fragment of 260 bp in 5% gels and 160 bp in 8% gels.
9.	Stop the electrophoresis and dismantle the apparatus leaving the gel adhering to one plate[c]. Cover the gel with thin plastic film and make an asymmetric pattern of radioactive marks around the gel by sticking pieces of tape marked with ^{32}P onto the cling film. Expose to X-ray film for 5−30 min. Use the autoradiogram to localise the labelled bands and cut out the slower-moving band or bands and electroelute the DNA as in *Table 5*.
10.	Recover the DNA by ethanol precipitation (*Table 5*, step 9). *Note*: since the DNA is now single-stranded it is essential to use siliconised tubes to achieve good recovery of the DNA.
11.	Dissolve the DNA in 10−50 μl of sterile water at a concentration of 1 fmol/μl (150 pg of a 500 nucleotide fragment is 1 fmol).

[a]From ref. 14.
[b]0.5 x TBE buffer: 45 mM Tris, 45 mM boric acid, 1.25 mM EDTA.
[c]This can be facilitated by coating one of the gel plates with Bind-Silane (LKB) and the other with a conventional siliconising fluid.

the hybridisation was performed in probe excess, the intensity of the band will be proportional to the concentration of the complementary RNA species in the RNA population. Thus the technique can also be used to compare the concentrations of individual RNA species in different populations.

If the probe was not completely strand-separated, some DNA-DNA duplexes will be present and these will be resistant to nuclease S1. Whether they will be detected on the final autoradiograph depends on the conditions of the experiment. It is not a problem for 5' end-labelled probes made by labelling a restriction site that leaves a 5'-protruding end using polynucleotide kinase since the nuclease S1 degrades the protruding nucleotides and thus removes the label from the re-annealed probe. For 3'-labelled probes, however, the label is not removed by the nuclease and the re-annealed probe can be seen on the final autoradiograph (see *Figure 2*).

Another problem to be aware of is that the nuclease S1 may cut at AT-rich regions in the probe causing spurious bands lower down the gel. This problem can be diagnosed by also running a 'minus RNA' control. If this is clean and there is a band of the size of the original probe then the probe has been completely protected by the RNA and therefore does not span the terminus or splice junction. A smear low down in the

Table 7. Hybridisation and Digestion with Nuclease S1.

1. Prepare the hybridisation mixture in a siliconised microcentrifuge tube as follows:
   DNA probe: 2.5 μl containing 2.5 fmol of labelled probe (i.e. 400 pg of a 500 nucleotide fragment).
   RNA: up to 5.5 μl in sterile water containing 0.25–0.5 fmol of RNA complementary to the probe
   (i.e., 0.1–0.2 ng of a 1 kb mRNA)[a],
   5 x hybridisation buffer (2 M NaCl, 50 mM Pipes, pH 6.4 with NaOH): 2 μl.
   Water: to 10 μl final volume.

2. Draw the mixture into a 20 μl or 25 μl glass microcapillary by touching one end of the capillary against the solution. Carefully invert the microcentrifuge tube while holding the capillary in the tube and allow the solution to move to the centre of the capillary.

3. Seal both ends of the capillary in a bunsen flame and attach a piece of water-resistant tape labelled with water-resistant marker ink.

4. Incubate at 60–65°C[b] for at least 6 h by immersing the capillaries in a water bath (an overnight incubation is permissible).

5. Just before use (in step 7 below) prepare 190 μl of nuclease S1 mixture for each hybridisation reaction to be analysed. To do this, mix:
   20 μl 10 x S1 digestion buffer[c]
   20 μl nuclease S1 (2000 U)[d]
   150 μl water

6. Rinse and wipe the outside of each capillary tube. Snip off 0.5 cm at each end using a glass cutter.

7. Expel the contents of each tube into 190 μl of nuclease S1 mixture (prepared in step 5) in a siliconised 1.5 ml microcentrifuge tube. Incubate at 37°C for 30 min.

8. Add 5 μl of tRNA (1 mg/ml) and recover the DNA by ethanol precipitation (*Table 5*, step 9).

9. Dry the DNA under vacuum and dissolve it in 10 μl of formamide dye mix (80% formamide[e], 0.5 x TBE buffer[f], 0.02% xylene cyanol). Heat at 90°C for 2 min. Chill on ice. Analyse on a denaturing 5% or 8% polyacrylamide gel containing 8 M urea and run in 0.5 x TBE buffer (e.g., see ref. 8). Also run suitable size markers.

[a]When the proportion of the complementary RNA in the total RNA population is not known, a range of RNA concentrations should be used.
[b]For a hybrid which is very short, or which has a high content of AT base pairs, it may be necessary to reduce the annealing temperature — see also footnote to *Table 9*.
[c]10 x S1 digestion buffer: 2 M NaCl, 20 mM $ZnSO_4$ 0.3 M sodium acetate, pH 4.6. Store at −20°C.
[d]Store the nuclease S1 at 10^5 U/ml in 0.4 M NaCl, 50% glycerol, 20 mM sodium phosphate buffer, pH 6.9, at −20°C.
[e]See Chapter 5, *Table 5*, footnote (c) for de-ionisation of formamide.
[f]See *Table 6*, footnote (b) for composition.

gel indicates that the RNA was degraded.

Finally, multiple bands are often seen. Although this may be due to heterogeneity at the 5′ or 3′ ends of the RNA, it is more likely to be due to exonucleolytic cleavage of the DNA-RNA hybrid by the S1 enzyme (often called 'nibbling'). Thus it is often necessary to perform the nuclease S1 digestion using different amounts of enzyme and different temperatures of incubation to determine the likely end point. *Figure 2* illustrates the 'nibbling' problem. In this particular case, the sequence in the area of the gene near the polyadenylation site for mRNA contains a high proportion of AT base pairs. This leads to 'breathing' of the hybrid at this point and consequent 'nibbling' by the nuclease S1. In an attempt to control for this, the nuclease S1 digestion (*Figure 2*) was performed under the standard (37°C) conditions and at 20°C, 30°C and 42°C. At the two lower temperatures a higher proportion of the probe yields protected fragments which are *longer* than expected for the correct 3′ end, indicating that digestion by nuclease S1 is incomplete. At 42°C the proportion of longer transcripts is greatly reduced

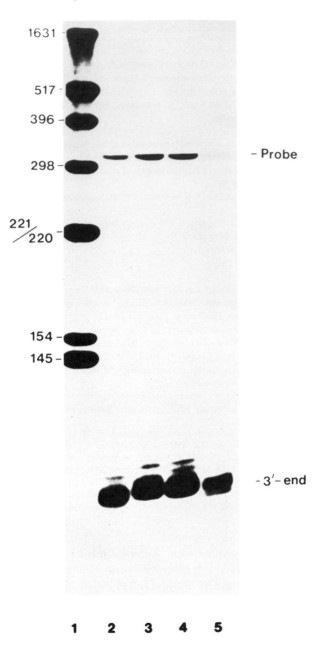

Figure 2. Identification of the 3′ terminus of an mRNA by nuclease S1 mapping; the effect of varying digestion conditions. In this experiment a 316-nucleotide fragment from the β1 globin gene of *Xenopus laevis* was hybridised to erythrocyte poly(A)$^+$ RNA. The probe was a 3′ end-labelled restriction fragment prepared by digestion with *Hind*III and *Hinf*I followed by strand separation on a native gel. There was a small amount of cross-contamination of the two strands which accounts for the presence of some renatured probe. Because this is a poly(A)$^+$ mRNA the length of the protected fragment indicates the position of the site of polyadenylation. Nuclease S1 digestion was performed at 37°C (**lane 2**), 20°C (**lane 3**), 30°C (**lane 4**) and 42°C (**lane 5**). **Lane 1** is a *Hing*I digest of pAT153 run as a size marker.

but the proportion of a fragment *smaller* than expected for the correct 3′ end is increased. At all temperatures, however, there are multiple fragments in the approximate position expected for DNA protected by the poly(A)$^+$ RNA. Thus, changing the temperature of nuclease S1 digestion has not reduced the 'nibbling' effect. This experiment nicely illustrates the limitation of nuclease S1 mapping, that is, it is normally very difficult to define a precise 'end point' for the nuclease S1 digestion. Almost inevitably, therefore, the precise position of an interruption or terminus can only be defined to within a few nucleotides. Therefore if it is important to define a 5′ terminus precisely for an RNA, both S1 mapping and primer extension should be used since the artefacts observed with the two methods are quite different.

4.3 Double-stranded Probes

With some DNA fragments it is not possible to separate the strands by the simple procedure given above. In this case, S1 mapping must be carried out with a double-stranded probe. In order to prevent the DNA probe re-annealing, the hybridisation is performed in a high formamide concentration (*Table 8*). Under these conditions, DNA-RNA hybrids are more stable than DNA-DNA duplexes. Hence it is possible to select a hybridisation temperature which allows hybridisation of the probe to the RNA but which prevents re-annealing of the probe. The optimum temperature depends on length of the hybrid, the G-C content and the primary sequence. Therefore when using a probe for the first time, it is necessary to try a range of temperatures around the expected optimum (see *Table 8* footnote). The precise conditions of hybridisation (i.e., salt and formamide concentrations) are crucial. Therefore to prevent any change in these by evaporation, the hybridisation is carried out in sealed capillaries. Rapid transfer after denaturation

Table 8. High Formamide Hybridisation for Double-stranded Probes.

1.	Into a siliconised microcentrifuge tube, place the double-stranded probe and the RNA in the amounts given in *Table 7*, step 1.
2.	Co-precipitate them with ethanol (*Table 5*, step 9), wash the pellet with 70% ethanol and dry it.
3.	Resuspend the pellet in 10 μl 80% formamide[a], 0.4 M NaCl, 1 mM EDTA, 50 mM Pipes, pH 6.4. Vigorous vortexing may be needed to completely dissolve the nucleic acids. Their dissolution can be checked using a Geiger counter.
4.	Draw the solution up into a 20 μl or 25 μl glass capillary. Seal it and mark it as described in *Table 7* (step 3)
5.	Immerse the capillaries in a 75°C water bath for 5 min to denature the DNA.
6.	Transfer the capillaries rapidly to a water bath set at the hybridisation temperature[b] and incubate for at least 6 h.
7.	Snap freeze the reactions by transferring the capillaries rapidly to a tray of dry ice.
8.	Remove the capillaries one by one from the dry ice. Snip off 0.5 cm at each end, wipe the outside and expel the contents into 190 μl of ice-cold nuclease S1 digestion mixture (*Table 7*, steps 5 − 7). Vortex each and stand on ice until all the tubes are ready.
9.	The rest of the procedure is identical to that described in *Table 7* (steps 8 and 9) for single-stranded probes.

[a]See Chapter 5, *Table 5*, footnote (c) for de-ionisation.
[b]The optimal hybridisation temperature varies with the base composition and length of the hybrid (for hybrids of <200 nucleotides) To determine the optimum, in an initial experiment vary the temperature in 2°C steps around the expected optimum of 50°C for a hybrid containing 50% GC base pairs.

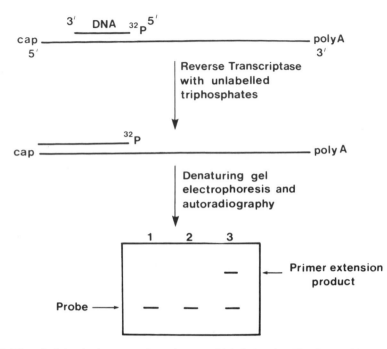

Figure 3. The principle of primer-extension using an end-labelled probe. The diagram illustrates how the 5' terminus of a polyadenylated mRNA might be determined using a single-stranded DNA probe. Normally the probe would be a fragment derived from a cloned copy of the gene by cleavage with appropriate restriction enzymes. The schematic representation of the autoradiogram (lower part of figure) demonstrates the results expected from a typical experiment. **Lane 1**, untreated probe; **lanes 2** and **3**, probe incubated under annealing conditions in the absence or presence of complementary RNA, respectively, and then incubated with reverse transcriptase in the presence of unlabelled triphosphates. Only when complementary RNA is present to form a hybrid with the primer is an extension product synthesised (**lane 3**).

and rapid freezing after hybridisation ensure that the reaction is not subjected to temperatures below the optimum, when probe re-annealing would be favoured.

5. PRIMER EXTENSION

5.1 **General Comments**

The principle of this technique is illustrated in *Figure 3*. It is the direct converse of nuclease S1 mapping. A radioactively-labelled probe derived entirely from within the gene is hybridised to complementary RNA and extended using the enzyme reverse transcriptase. The cloned probe is normally derived from a region near the 5' end of the gene and the extension reaction terminates at the extreme 5' end of the RNA. As with S1 mapping, this technique can be used to determine precisely the start point of transcription of an mRNA sequence.

The primer extension reaction is more sensitive and yields cleaner results using poly(A)$^+$ RNA. With total RNA there are a relatively large number of prematurely terminated transcripts (i.e., there are many primer extension products which are longer than the probe but which do not extend to the true 5' terminus of the mRNA). These presumably result from an inhibitor of reverse transcriptase which is removed by oligo(dT)-cellulose chromatography [the normal method of purification for poly(A)$^+$ RNA]. However, with a probe derived from near the 5' end of the mRNA (i.e., within ~200 nucleotides) perfectly acceptable results can certainly be obtained using total RNA so poly(A)$^+$ RNA need only be prepared if the level of prematurely terminated transcription is unacceptable.

5.2 Choice of Probe

The radiolabelled probe may be either single-stranded or double-stranded DNA but, as with S1 mapping, there are great advantages to be gained from using a single-stranded probe whenever possible. However, because only a small fragment of DNA is required for primer extension, a simpler and more generally applicable strand separation procedure can be utilised (13). The general approach is as follows. Two restriction enzymes which produce cohesive ends of different lengths are utilised to generate a restriction fragment from near the 5' terminus. Because they will differ in length the coding strand can be separated from the non-coding strand by electrophoresis on a denaturing polyacrylamide gel. Since electrophoretic resolution of sequences decreases with increasing fragment size, there is a practical upper limit of about 100 nucleotides to the length of the probe. Thus a pair of restriction enzyme cleavage sites should be chosen which are separated by between 20 and 100 nucleotides, and which lie within 200 nucleotides of the expected 5' end of the gene. In fact, it is not absolutely imperative that the probe be derived from a region near the 5' terminus but there will be fewer prematurely-terminated transcripts with such a probe and the resolution of the transcripts of multigene families (e.g., see *Figure 4*) will be increased.

Normally, two enzymes which cut the recombinant molecule infrequently should be used. Otherwise the large number of fragments which are generated reduces the efficiency of the labelling reaction and renders fragment isolation much more difficult. However, if only one enzyme which cleaves the recombinant infrequently can be found, then this enzyme should be used alone to generate a free end for labelling. The second, more frequently cutting enzyme will then generate a large number of unlabelled fragments but these will not interfere with the procedure. Finally, if the only suitable enzymes cut frequently (i.e., more than two or three times) within the recombinant, then it is necessary to purify a suitable fragment by gel electrophoresis under non-denaturing conditions (see Section 4.2.2 and ref. 8.).

When two enzymes cannot be found which generate cohesive ends of different length, it is almost always possible to generate a difference in strand length by filling in one of the 3' ends using the Klenow fragment of *E. coli* DNA polymerase. This is normally achieved by cutting with the enzyme which cleaves nearer to the 5' terminus of the gene, filling in the 3' ends of the fragment and then cleaving with the second enzyme. By using only a selected sub-set of the four triphosphates in the filling in reaction the number of potential strand-separation strategies can be increased. Although this procedure will allow the 5' terminus of the gene to be determined, its drawback is that a 3'-labelled probe is generated, and so the sequence of the primer-extension product cannot be determined subsequently using the chemical degradation procedure of Maxam and Gilbert (14).

153

5.3 Choice of Labelling Method

The choice of labelling method depends in part on the aim of the experiment and in part on the strategy used to obtain strand separation.

(i)   *5' end-labelling.* This method must be used if the sequence of the primer extension products is to be determined.

(ii)   *3' end-labelling.* This method is satisfactory if the aim of the experiment is simply to locate the position of 5' termini or to quantitate the amount of a particular primer extension product.

(iii) *Uniform labelling.* A uniformly labelled probe may be prepared by synthesis on a single-stranded template or by replacement synthesis. This generates a probe of very high specific activity which will give a very high level of sensitivity if used before significant strand scission resulting from radioactive decay has occurred (15).

Detailed protocols for these labelling procedures are described in Chapter 2. It is important to know the specific activity of the probe so that the correct amount of DNA can be used in the hybridisation. This should be calculated immediately after the labelling reaction by determining the number of Cerenkov counts which are excluded from the column used to remove unincorporated nucleoside triphosphates. Note that this calculation assumes that *all* of the labelled nucleotide has been incorporated in the DNA ends. Hence when labelling 5' ends it is important to remove contaminating RNA from plasmid preparations. After recovery of the single-stranded DNA from the denaturing gel used for strand separation, dissolve the radioactively-labelled probe in water at a concentration of 1 fmol per μl (15 pg of a 50 nucleotide primer is 1 fmol). For a 5'- or 3'-labelled fragment the specific activity should be approximately 5×10^6 d.p.m./pmol of primer.

5.4 The Hybridisation and Primer Extension Reactions

The procedures for hybridisation and primer extension are described in *Table 9*. It is necessary to exercise care in optimising the conditions under which the probe is hybridised to the RNA. The hybridisation reaction is performed under conditions of moderate (5- to 10-fold) excess of DNA over complementary RNA. While it is important to ensure that an adequate DNA excess is achieved, a massive excess of primer should not be used because this will lead to non-specific priming during the extension reaction. It is therefore wise to perform an RNA titration around the expected optimum.

The optimal temperature for the hybridisation reaction using a primer of between 20 and 100 nucleotides will be between 50°C and 70°C. *Table 10* gives optimum temperatures for a number of primers and can be used to predict the likely optimum. However in initial experiments it is best to perform parallel annealing reactions at different temperatures around the expected optimum (from 10°C below to 10°C above in 5°C steps). This will also greatly aid the interpretation of the data obtained. At the end of the annealing reaction the hybridisation mixture is diluted to lower the salt concentration and the primer is then extended using reverse transcriptase and unlabelled nucleoside triphosphates. This is performed at 42°C to minimise secondary structure in the mRNA which can lead to premature termination of the reverse transcriptase reaction. The extension products are then separated on a high resolution urea-acrylamide gel of suitable composition.

Table 9. Procedure for Hybridisation and Primer Extension.

1.	In a plastic microcentrifuge tube, mix: DNA probe: 2.5 μl containing 2.5 fmol of primer (i.e., 40 pg of a 50 nucleotide primer) RNA: up to 5.5 μl in sterile water containing 0.25−0.5 fmol of RNA complementary to the probe (i.e., 0.1−0.2 ng of a 1 kb mRNA[a]). 5 x Hybridisation buffer (2 M NaCl, 50 mM Pipes, pH 6.4): 2 μl Water: to 10 μl final volume.
2.	Seal the reaction mixture in a 20 or 25 μl glass capillary (as described in *Table 7*, step 3) and anneal for 6 h at an appropriate temperature[b].
3.	Break open the capillaries (as in *Table 7*, step 6) and expel the contents of each into 90 μl of extension reaction buffer. This reaction buffer contains (per 90 μl): 5 μl of 1 M Tris pH 8.2 5 μl of 0.2 M DTT 5 μl of 0.12 M magnesium chloride 2.5 μl of actinomycin D (1 mg/ml in water)[c] 5 μl of 10 mM dCTP 5 μl of 10 mM dATP 5 μl of 10 mM dGTP 5 μl of 10 mM dTTP 10 units of reverse transcriptase[d] water to 90 μl final volume
4.	Incubate the primer extension reaction for 1 h at 42°C.
5.	Recover the nucleic acid by ethanol precipitation (*Table 5*, step 9).
6.	Centrifuge at 12 000 *g* for 5 min.
7.	Rinse the pellet with 0.5 ml of 95% ethanol. Then dry it for 1 min under vacuum.
8.	Dissolve the pellet in 5 μl of formamide dyes and electrophorese on a denaturing acrylamide gel alongside suitable labelled size markers (*Table 7*, step 9).
9.	Perform autoradiography of the gel using an intensifying screen.

[a]When the proportion of the complementary RNA in the total RNA population is not known, a range of RNA concentrations should be used.
[b]Determination of an appropriate hybridisation temperature; see the text and *Table 10*.
[c]Actinomycin D is included to prevent self-copying of the primer by reverse transcriptase.
[d]Reverse transcriptase obtained from different manufacturers varies greatly in activity per unit. The figure of 10 units is based on enzyme obtained from Life Sciences Inc. When using a new batch of enzyme it is advisable to perform a titration in which replicate hybridisation reactions are subjected to primer extension using various amounts of enzyme around the expected optimum.

Table 10. Optimal Hybridisation Temperatures.

This Table gives experimentally determined optimal hybridisation temperatures for a number of single-stranded probes of different base composition and length. Using these data the approximate optimum temperature for hybridisation of a given probe can be predicted but it is advisable to perform a preliminary experiment using a range of temperatures (see the text).

Length of primer (nucleotides)	Base composition (%GC)	T_{opt} (°C)
29	38	55
38	37	55
42	38	60
42	60	60
57	46	60
52	52	65

5.5 **Interpretation of Primer Extension Data**

Even with a probe at the lower end of the specific activity range specified above, an RNA constituting only 0.1% of the population can easily be detected in an overnight exposure of the gel. It is advisable when deciding upon electrophoresis conditions to ensure that the unhybridised primer does not run off the gel. Because the hybridisation goes to completion under the conditions specified (J.G.Williams and H.Mahbubani, unpublished results), and because there is an excess of primer, it is possible to gauge the concentration of the complementary RNA by estimating the proportion of primer which has been extended.

The primer extension reaction will often yield more than one transcript. There are several possible reasons for this.

(i)   *'Cap effects'*. Generally a full length reverse transcript of a eukaryotic mRNA sequence will also be accompanied by a minor reverse transcript which will migrate during electrophoresis with an apparent length about one to two nucleotides shorter than the full length transcript. This is thought to be due to premature termination of reverse transcription at the methylated residue situated next to the cap site.

(ii)   *Multiple start sites of transcription*. Many genes which are transcribed by RNA polymerase II have multiple start sites for transcription which will yield mRNA sequences differing in length in their 5'-non-coding region.

(iii)   *Cross-hybridisation to related RNA transcripts*. The different transcripts from multigene families may share sufficient homology over the region of the primer used to yield multiple reverse transcripts in the primer extension reaction, each derived from a different mRNA.

(iv)   *Premature termination of reverse transcription*. This can be a problem when using a primer which hybridises to a region distant from the 5' terminus and it is amplified if total rather than poly(A)$^+$ RNA is utilised. It may be caused by regions of secondary structure in the RNA which prevent copying by reverse transcriptase or by cleavage of the RNA in particularly sensitive regions by contaminating RNase.

Figure 4 shows typical results obtained when the primer extension technique is used to analyse the transcripts of a multigene family. In this experiment a probe prepared from a conserved region of a *Dictyostelium* actin gene was hybridised to poly(A)$^+$ RNA. Seven transcripts were reproducibly obtained at the optimal hybridisation temperature (defined as the temperature at which the maximum number of transcripts were obtained − in this case 60°C). The same result was also obtained with total RNA. Reduction of the hybridisation temperature produced a uniform decrease in the amounts of the various transcripts. However, as the temperature was increased, there was a differential loss of transcription products (e.g., 1a, 1b, 2, 4) until at the highest temperature only three transcripts were obtained (3a, 3b and 3c). These three transcripts are derived from the same gene but they differ in the length of their 5'-non-coding regions. The transcripts which were lost as the temperature was elevated are copied from other actin mRNA sequences which cross-hybridise to this probe. Nucleotide sequence analysis of preparative primer extension reaction products (using a 5' end-labelled probe) showed

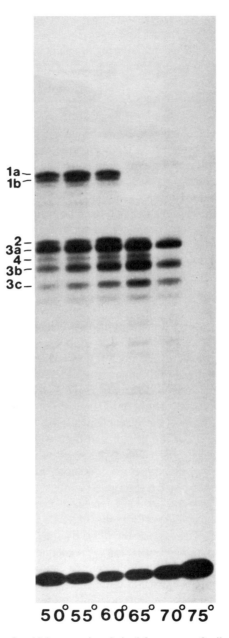

Figure 4. The identification of multiple transcripts derived from a gene family by primer extension; the effect of varying the hybridisation temperature. In this experiment a 42-nucleotide fragment of an actin gene from *Dictyostelium discoideum* was hybridised to *Dictyostelium* total poly(A)$^+$ RNA (15). The probe was a 3' end-labelled restriction fragment prepared by sequential digestion with *Hind*III and *Hpa*II. The fragment was end-labelled with the Klenow fragment of DNA polymerase I after restriction with *Hind*III but before *Hpa*II cleavage. This generated a 2-nucleotide difference in the length of the coding and anti-coding strands permitting their separation on a denaturing acrylamide gel. The fragment was hybridised to the RNA at the temperatures indicated. The various reverse transcripts (1a, 1b, 2, 3a, 3b, 3c and 4) are described in the text.

that the seven transcripts are derived from four different actin genes (15). Two of the genes (1 and 3) display 5'-terminal heterogeneity, that is, they give rise to multiple transcripts (1a and 1b; 3a, 3b and 3c) differing in length at their extreme 5' ends because of heterogeneity in the precise position at which transcription is initiated.

5.6 Large-scale Procedure for Sequence Determination

As described in the previous section in relation to *Figure 2*, the great advantage of the primer extension technique is that the relationship between various transcripts may be determined directly by establishing the nucleotide sequence of the primer extension products (15). This requires that the reaction be scaled up to yield sufficient radioactivity in the reverse transcript for sequencing to be possible. The minimum amount of radioactivity required for DNA sequence analysis by the Maxam-Gilbert procedure (14) is about 2000 c.p.m. Thus the reaction must be scaled up such that the least abundant transcript, for which a sequence is to be determined, yields this amount of radioactivity. At levels lower than this, useful information can still be obtained by running just one of the four reactions (e.g., the 'G+A' reaction) since this will give some indication of the relationship between the various transcripts.

The scale of the preparative reaction required is determined by the abundance of the RNA and the specific activity of the probe. After autoradiography of the gel used for the small-scale analysis described above, excise each primer extension product from the gel and determine the amount of radioactivity in the slice using Cerenkov counting.

The volume of the primer extension reaction will be dictated by the amount of poly(A)$^+$ RNA used in the hybridisation. If a large amount of RNA (>100 μg) has to be used, the volume of the reaction must be increased because the upper solubility limit of RNA is approximately 10 mg/ml. The ratio of RNA to DNA may be varied to make the most efficient use of the available materials. It is permissible to use a molar excess of RNA to drive all of the primer into hybrid. When using large amounts of primer and RNA in a small final volume, the reaction will be complete within a few minutes so that a hybridisation period of 1 h is more than adequate. In order to avoid wasting a large amount of reverse transcriptase unnecessarily in a large-scale reaction, it is advisable to remove several small aliquots (each 1 or 2% of the total volume) from the hybridisation reaction and to perform pilot extension reactions using various levels of reverse transcriptase. The bulk of the hybridisation mixture can be stored frozen at $-20°$C. Once an enzyme level sufficient to give a plateau level of extension product has been determined, the remainder of the sample can be processed after dilution into a suitable volume of primer extension buffer.

There is one additional step which is normally required for preparative scale reactions. If more than about 40 μg of RNA is used in the hybridisation the RNA must be degraded to prevent smearing during electrophoresis. This procedure is described in *Table 11*.

After RNA digestion, gel electrophoresis is performed as for the small-scale procedure (*Table 9*). Finally the primer extension products are recovered from the gel and their sequence determined by the chemical degradation procedure of Maxam and Gilbert (14).

Table 11. Degradation of RNA in the Large-scale Procedure.

1. Add sodium acetate (pH 7.0) to a final concentration of 0.3 M.
2. Add heat-treated[a] pancreatic RNase to a final concentration of 25 μg/ml.
3. Incubate for 15 min at room temperature.
4. Extract with an equal volume of re-distilled phenol.
5. Centrifuge briefly to separate the phases. Remove the aqueous phase to a new tube, re-extract the organic phase with 50 μl of 0.3 M sodium acetate (pH 7.0) and add to the first aqueous phase.
6. Add two volumes of ethanol and then recover the RNA transcripts for gel electrophoresis as described for the small-scale procedure (*Table 9*, steps 6−9).

[a]RNase is treated by placing in a boiling water bath for 10 min before use.

Table 12. Determination of the Proportion of the Primer in Hybrid Form Using Nuclease S1.

1. At the end of the hybridisation reaction, remove 2 μl of the mixture into 210 μl of nuclease S1 mixture (*Table 7*, step 5).
2. Place 4 x 50 μl aliquots into microcentrifuge tubes.
3. Add 300 units of nuclease S1 (3 μl at 10^5 units/ml) to only two of the aliquots.
4. Incubate all four tubes for 30 min at 42°C.
5. Remove 45 μl from each tube and spot it onto a disc of DEAE paper (DE81; Whatman) which has been labelled with a soft lead pencil.
6. Place the four discs in a 1 litre beaker containing 200 ml of 5% Na_2HPO_4 (prepared by adding 50 g of solid Na_2HPO_4 to 1 litre of water stirred during the addition with a magnetic stirring bar).
7. Swirl the beaker gently for about 1 min. Pour off the buffer.
8. Repeat this washing step four times more.
9. Wash the filters twice with tap water, once with ethanol and then air-dry the filters.
10. Count the filters in a suitable scintillant. Determine the proportion of the probe which is resistant to nuclease S1 by averaging the duplicate values and expressing the radioactivity of the samples with nuclease S1 as a percentage of the radioactivity of the samples with no nuclease S1.

5.7 Potential Problems with Primer Extension

The procedure described in this chapter has proved to be highly reproducible using RNA from diverse sources and purified by a number of different procedures. The only step which may give problems is the reverse transcriptase reaction, either because too little enzyme is used or because an inhibitor of reverse transcriptase has been co-purified with the primer or with the RNA. The simplest way to test for this problem is to determine whether a reasonable proportion of the DNA primer has hybridised to RNA in the hybridisation step. This is described in *Table 12*. If the reverse transcriptase is working correctly then *all* of the primer which has formed a hybrid with the RNA should be extended. It may be difficult to quantitate this precisely in situations where there is a very large excess of DNA primer over RNA because only a small proportion of the primer will participate in hybrid formation. It may therefore be necessary to perform this analysis at several different ratios of DNA to RNA such that at the highest ratio of RNA a significant proportion of the primer forms a hybrid.

6. REFERENCES

1. Clemens,M.J. (1984) in *Transcription and Translation − A Practical Approach*, Hames,B.D. and Higgins,S.J. (eds.), IRL Press, Oxford and Washington DC, p. 211.
2. McMaster,G.K. and Carmichael,G.G. (1977) *Proc. Natl. Acad. Sci. USA*, **74**, 4835.

3. Lehrach,H.D., Diamond,J.M. and Boedtker, (1977) *Biochemistry (Wash.)*, **16**, 4743.
4. Bailey,J.M. and Davidson,N. (1976) *Anal. Biochem.*, **70**, 75.
5. Alwine,J.C., Kemp,D.J., Parker,B.A., Reiser,J., Renart,J., Stark,G.R. and Wahl,G.M. (1979) in *Methods in Enzymology*, Vol. **68**, Wu,R. (ed.), Academic Press Inc., London and New York, p. 220.
6. Thomas,P.S. (1980) *Proc. Natl. Acad. Sci. USA*, **77**, 5201.
7. Thomas,P.S. (1983) in *Methods in Enzymology*, vol. **100**, Wu.R., Grossman,L. and Moldave,K. (eds.), Academic Press Inc., London and New York, p. 255.
8. Sealey,P.G. and Southern,E.M. (1982) in *Gel Electrophoresis of Nucleic Acids — A Practical Approach*, Rickwood,D. and Hames,B.D. (eds.), IRL Press, Oxford and Washington DC, p. 39.
9. Wahl,G.M., Stern,M. and Stark,G.R. (1979) *Proc. Natl. Acad. Sci. USA*, **76**, 3683.
10. Holmes,D.S. and Quigley,M. (1981) *Anal. Biochem.*, **114**, 193.
11. Berk,A.J. and Sharp,P.A. (1977) *Cell*, **12**, 721.
12. Weaver,R.F. and Weissman,C. (1979) *Nucleic Acids Res.*, **7**, 1175.
13. Proudfoot,N.J., Shander,M.H., Manley,J.L., Gefter,M.L. and Maniatis,T. (1980) *Science (Wash.)*, **209**, 1329-1336.
14. Maxam,A. and Gilbert,W. (1980) in *Methods in Enzymology*, vol. **65**, Grossman,L. and Moldave,K. (eds.), Academic Press Inc., London and New York, pp. 65.
15. Tsang,A.S., Mahbubani,H. and Williams,J.G. (1982) *Cell*, **31**, 375.

CHAPTER 7

Electron Microscopic Visualisation of Nucleic Acid Hybrids

PIERRE OUDET and CHRISTIAN SCHATZ

1. INTRODUCTION

In 1959, using the properties of a protein film made at an air-water interphase, Kleinschmidt and Zahn first described spreading techniques allowing the direct visualisation of double-stranded DNA filaments (1). Since then, this approach has been further developed so that it is now possible to characterise single-stranded DNA and RNA and hybrid molecules. The reproducibility of this technique, as well as the use of internal standards, allows the researcher to obtain quantitative information and to map precisely the different structures observed. These developments have become even more powerful and useful with the advent of the new techniques of molecular biology, that is, purification of restriction fragments and recombinant DNA technology.

Electron microscopy of nucleic acids is a tool of increasing interest in molecular biology (for reviews, see refs. 2−5). Among the specific results that can be obtained from the visualisation of nucleic acids are:

(i)   the characterisation of DNA structure (circular or linear form, degree of super-coiling, length distribution of DNA fragments);
(ii)   the delineation of A+T-rich regions in DNA molecules by partial denaturation;
(iii)   the analysis of complex structures such as replicative intermediates;
(iv)   the detection and analysis of complementary sequences either between two DNA populations (heteroduplexes) or between an RNA and a DNA molecule (hybrids) by D-loop mapping or R-loop and R-hybrid mapping, respectively;
(v)   the study of protein-nucleic acid interactions.

The purpose of this chapter is to describe practical approaches to the analysis of nucleic acid hybrids using electron microscopy. In practice, three different steps are required:

(i)   denaturation of the nucleic acids followed by renaturation under the desired conditions;
(ii)   spreading of the hybrids in the presence of proteins and adsorption on carbon films prepared on electron microscope grids;
(iii)   visualisation and length measurement of the hybrid molecules either directly on a video monitor connected to the electron microscope or on prints obtained from electron microscopy photographic negatives.

These techniques are described in this chapter in the form of protocols currently in use in the authors' laboratory. With slight modifications, depending on the molecules involved, the same techniques can be used for the characterisation of many different types of molecules. These include DNA-DNA heteroduplexes such as between cDNA and genomic DNA fragments, to examine the homology between the two DNA sequences, and DNA-RNA hybrids such as genomic DNA-mRNA for mapping the coding sequences and the orientation of transcription. However, it should be noted that the use of the electron microscope for the analysis of hybrid molecules requires a great deal of practical experience before being routinely employed in a laboratory. Operators need to practice the methodology several times before having enough experience and control at each step to ensure reasonable reproducibility and reliability of the resulting data.

2. HYBRID FORMATION

The visualisation by electron microscopy of hybrids between two populations of DNA molecules (heteroduplexes) or between DNA and RNA provides reliable information for the characterisation of partial or total homology between the two populations of nucleic acid molecules.

2.1 DNA-DNA Hybridisation

The principle of this method consists of denaturating the two DNA populations, followed by slow annealing of the mixed populations and then spreading of the mixture under conditions which allow the operator to discriminate between single- and double-stranded regions of the hybrid molecules by electron microscopy and to measure their lengths (6). Regions of homology will be double-stranded (thick filaments, well extended) and regions lacking homology will be single-stranded segments (thinner filaments) or 'D-loops'.

2.1.1 *Basic Procedure*

Originally, high temperature (7) or alkaline pH (6) were used to either totally or partially denature DNA duplexes. Formaldehyde was then used to fix the denatured regions. However, the presence of formamide, which allows denaturation at much lower temperatures and avoids extreme pH, improves the reproducibility and the quality of the preparations. In addition, the presence of formamide in the spreading solution improves the characterisation of single-stranded regions (8).

The method used routinely in the authors' laboratory (7) is based on the technique described by Wellauer *et al.* for the analysis of ribosomal genes from *Xenopus laevis* (9). Heteroduplex formation requires linearised DNA molecules. Therefore, if the source of DNA is a circular supercoiled recombinant, it must be linearised using a suitable restriction enzyme before use.

(i)   Immediately before each hybridisation experiment, prepare the denaturation mixture in a 1.5 ml microcentrifuge tube by mixing:

       7 μl of deionised formamide (see Section 3.2.1.)

       1 μl of 10 mM EDTA, 3 M NaCl, 0.1 M Tris-HCl, pH 8.5.

(ii)   Add 1 μl of each of the two (linearised) DNA preparations (each at $10-30$ μg/ml in 0.2 mM EDTA, 10 mM Tris-HCl, pH 7.5). Mix gently.

(iii)   Completely denature the DNA molecules by incubating at 75°C for 5 min.

(iv)   Cool the sample immediately on ice. Leave for 5 min.

(v)   Hybridise by incubation at 25°C (in a water bath) for 30–60 min.

(vi)   Place the sample on ice until spreading. It is possible to keep an aliquot of the renatured DNA sample for some days at 4°C. This aliquot can be used later if the spreading has to be repeated.

2.1.2 *Variation in the Conditions of DNA-DNA Hybridisation*

The stringency of the hybridisation can be modified by varying the formamide concentration or the ionic strength and the temperature. The effect of variation of temperature and ionic conditions have been discussed extensively elsewhere (refs. 3,6; see also Chapter 3).

Formamide concentration and hybridisation temperature. A linear relationship exists between the formamide concentration and the melting temperature (T_m) of DNA molecules (8). Thus a 1% increase in the formamide concentration lowers the melting point of native duplex DNA by 0.72°C (10). The optimum conditions for hybrid formation depend on the G+C content of the molecules and on the distribution of the G+C base pairs along the DNA molecule. For example, the optimum temperature for renaturation of λ phage DNA is T_m −30°C (i.e., 25°C) at the formamide concentration (70%) given in the basic procedure above.

Hybridisation time. In most cases, about 50% of DNA molecules are renatured during the 30–60 min incubation recommended. After longer incubation times, more intramolecular aggregates generally appear (1,3–6,11). Therefore 30–60 min is the optimum time for general usage of the technique.

DNA concentration. The final concentration of each DNA preparation must be adjusted with regard to the molecular weight of the DNA. For two DNA populations of similar molecular weight, the optimum DNA-DNA ratio seems to correspond to identical concentration by weight. If the two species are different in size, the final concentration of the smaller species must be increased. For example, it has been found that if one DNA molecule is five times shorter than the other, its final concentration should be arranged to be twice the concentration of the longer one. In addition, the concentration of the shorter fragment should also be increased for good heteroduplex formation when the regions of homology within the DNA molecules are short with respect to the overall size of those molecules.

Integrity of DNA molecules. The presence of nicks in DNA molecules produces complex hybrids involving more than two DNA strands, introducing the formation of aggregates and branched structures due to the association of several molecules. Therefore, in order to be able to interpret the structure of the hybrids, it is clearly necessary to work with intact molecules, avoiding internal nicks as far as possible. The major causes of nicks in recombinant DNA seem to be u.v. irradiation in the presence of ethidium bromide, vigorous mixing and shearing forces due to the use of narrow pipettes. Intactness of the DNA strands can be checked by nuclease S1 treatment, or by electron microscopy by spreading the sample immediately after the denaturation step.

2.2 DNA-RNA Hybridisation

Essentially two methods have been developed to map sequences along a DNA molecule which are complementary to a particular RNA. The first method is the R-loop technique in which an RNA molecule is hybridised to its complementary sequence in duplex DNA (12,13). The second method is the R-hybrid technique where RNA molecules hybridise to denatured single-stranded DNA molecules (14).

2.2.1 R-loop Method

This procedure takes advantage of the higher stability of DNA-RNA hybrids compared with DNA-DNA duplexes of similar sequences. As a result, under destabilising conditions in the presence of formamide and high temperature (12,13,15), a single-stranded RNA molecule can displace its DNA copy and hybridise to the complementary strand of a duplex DNA. The displaced DNA strand ('R-loop') serves to indicate the site of hybridisation of the RNA to the DNA. The method used in the authors' laboratory is that outlined by Thomas *et al.* (12).

Determination of the temperature for R-loop formation. Knowing the G+C content of the DNA molecule, the temperature for maximum R-loop formation can be derived from the following equation (12):

$$T_{max} = 81.5 + 0.50 \ (\% \ G+C) + 16.6 \log [Na^+] - 0.60 \ (\% \ formamide).$$

An alternative procedure is to determine the temperature at which half of the DNA molecules are completely denatured (strand separation temperature $=T_{ss}$). This temperature is determined experimentally as follows:

(i)   Mix:

       70 μl of deionised formamide (see Section 3.2.1)

       10 μl of 10mM EDTA, 3 M NaCl, 0.1 M Tris-HCl, pH 8.5

       10 μl of linearised DNA (10−30 μg/ml)

       10 μl of distilled water

(ii)   Denature for 15 min at different temperatures starting from 45°C and increasing progressively by steps of 1°C.

(iii)   Take an aliquot (5 μl) at each temperature and keep on ice until spreading.

(iv)   After spreading (Section 3.2), examine the samples by electron microscopy.

The temperature at which half of the DNA molecules are fully denatured corresponds to the optimum temperature for R-loop formation. In fact, the incubation temperature can be modified ± 1°C around the T_{ss} depending on the nature of the DNA sequences to improve the rate of R-loop formation.

R-loop mapping: experimental procedure. Once the optimum temperature for R-loop formation has been determined, R-loop mapping is carried out as follows:

(i)   Mix:

       7 μl of deionised formamide (Section 3.2.1)

       1 μl of 10 mM EDTA, 3 M NaCl, 0.1 M Tris-HCl, pH 8.5

       1 μl of DNA (10−30 μg/ml)

       1 μl of RNA (30−100 μg/ml).

(ii)   Incubate for 4−8 h at the optimum temperature.

(iii)   Keep on ice until spreading (Section 3.2.3).

2.2.2 *R-hybrid Method*

Holmes *et al.* (14) developed the R-hybrid technique by modification of the R-loop method. The DNA molecules are first denatured and then are renatured with RNA to form DNA-RNA hybrids. The following procedure (16) is used in the authors' laboratory.

Determination of the optimum temperature for R-hybrid formation. The optimum temperature for R-hybrid formation is empirically defined as the temperature at which 50% of the DNA molecules are renatured after 3 h incubation. This temperature is determined as follows:

(i)   Mix:
        70 μl deionised formamide (Section 3.2.1)
        10 μl of 10 mM EDTA, 3 M NaCl, 0.1 M Tris-HCl, pH 8.5
        10 μl of DNA (10−30 μg/ml)
        10 μl of distilled water.

(ii)   Heat at 75°C for 5 min to denature the DNA and then immediately put on ice. Leave for 5 min.

(iii)   Incubate the mixture for 3 h at different precisely-maintained temperatures (see below) to allow hybridisation to occur, then take an aliquot (5 μl) at each temperature, and keep on ice until spreading.

(iv)   After spreading (Section 3.2.3), examine the DNA molecules by electron microscopy.

The incubation temperature is determined step by step starting from 60°C (several degrees above the melting temperature under these conditions of hybridisation). The temperature is decreased in 1°C steps until 50% of the molecules appear as double-stranded renatured filaments by electron microscopy. Once the optimum temperature has been determined approximately, a more precise value may be obtained by 0.1°C variations around the optimum temperature.

R-hybrid mapping: experimental procedure.   This is carried out as follows:

(i)   In a microcentrifuge tube, mix:
        7 μl deionised formamide (Section 3.2.1)
        1 μl of 10 mM EDTA, 3 M NaCl, 0.1 M Tris-HCl, pH 8.5
        1 μl of linearised DNA at 10−30 μg/ml

(ii)   Incubate at 75°C for 5 min to denature the DNA, then immediately put on ice. Leave for 5 min.

(iii)   Add 1 μl of RNA solution (30−100 μg/ml) in distilled water to the denatured DNA and incubate for 3−4 h at the optimum temperature for R-hybrid formation.

(iv)   After the incubation, keep the sample on ice.

(v)   For analysis by electron microscopy, dilute 5 μl of this sample 10-fold in the hyperphase solution ready for spreading (Section 3.2.3).

2.2.3 *Variation in the Conditions of DNA-RNA Hybridisation*

Optimum formation of DNA-RNA hybrids depends primarily on the temperature and time period for hybridisation and on the nucleic acid concentration.

Temperature. DNA-RNA hybrids are more stable than DNA-DNA duplexes (15). The optimum temperature for DNA-RNA hybridisation is therefore slightly higher than the temperature at which 50% of the DNA molecules are denatured (R-loop) or renatured (R-hybrid). Significant increases in the yield of either DNA-RNA hybrid or R-loop structures can be obtained by varying the optimum temperature by up to ±0.5°C.

Hybridisation time. At the optimum temperature, a sufficient number of R-loops or R-hybrids are formed within 4−8 h or 3−4 h, respectively, for electron microscopic characterisation. After this time, only a few additional hybrids are formed. DNA-DNA hybrids and DNA-RNA hybrids can be kept at 4°C for some days without aggregation or modification of the appearance of the hybrid molecules.

Nucleic acid concentration. The RNA concentration specified in the protocols above (30−100 μg/ml) was that used originally for an mRNA present at about 1% abundance in the total poly(A)$^+$ mRNA population (14). This assumes that all the mRNA molecules are intact. If a significant fraction is not, the mRNA concentration must be increased proportionately. The DNA concentration used (10−30 μg/ml) should be adjusted according to its molecular weight. For long DNA molecules (≥50 000 bp), use the DNA at 10 μg/ml but increase the concentration for smaller DNA molecules.

3. SPREADING OF NUCLEIC ACIDS IN A PROTEIN FILM

Since the original description of spreading of nucleic acids in the presence of proteins (1), many improvements and modifications have been described to solve specific problems (2,6,11). In particular, spreading in the presence of formamide is used to obtain good resolution in the visualisation of single-stranded nucleic acids and to discriminate between single- and double-stranded regions in DNA heteroduplexes or DNA-RNA hybrids (8). In practice, spreading of nucleic acids is quite reproducible provided the conditions are controlled precisely. However, variations in parameters such as ionic strength and pH and the presence of detergents or proteins other than that used for spreading may lead to aggregation of nucleic acids and tangling (3,4). The following sections describe conditions that routinely give good results.

3.1 Preparation of Electron Microscope Grids coated with Supporting Film

3.1.1 *Preparation of the Grids*

Different types of grids can be used for electron microscopy. The choice of the type of grid (e.g., nickel or copper) and the mesh size depends on the nature and thickness of the support film and the length of the molecules to be studied. Usually the authors use 300−400 mesh copper grids (Fullam) to visualise nucleic acids 500−50 000 bp long. The grids are first washed and then covered with a thin carbon film before use. To do this, place 200−500 grids in a glass tube and wash them twice for 30 min each in dichloro-1,2-ethane, agitating the solution using a Pasteur pipette. Then rinse the grids in absolute ethanol. Finally, remove the grids from the glass tube and dry them on a filter paper placed in a Petri dish. They are then ready to be covered by the carbon supporting film.

3.1.2 *Preparation of the Supporting Film*

The quality of the spreading and contrast of the nucleic acids in the electron microscope is very much dependent on the thinness and homogeneity of the supporting film. Usually support films are prepared either from pure carbon film on mica sheets (as described below), or from a collodion or formvar solution which is then carbon coated (17). The carbon film has the advantage of being thin, stable under the electron beam and sufficiently hydrophobic to allow a good adsorption of the protein film. The disadvantage is that a film which is too thin may break in the electron microscope, so the thickness has to be adjusted to the size of the holes (i.e., gauge of the grid) over which it is deposited. Very thin carbon film ($<3-10$ nm) can be used when stabilised on thick 'perforated' films (18).

To prepare pure carbon film:

(i)   Carefully cleave mica sheets (4 cm x 4 cm; Pelanne Instruments) using forceps.

(ii) Immediately place them in an evaporator, placing the freshly cleaved, clean surface towards the carbon source.

(iii) Evaporate the carbon from an electron beam device (Edwards) at a vacuum of $10^{-6}-10^{-7}$ Torr (*Figure 1*). The better the vacuum, the cleaner and more homogeneous will be the film. Place the mica sheet at a distance of 20 cm from the source and carefully control the extent of the vaporisation by slow and progressive increase of the voltage. The thickness can be checked optically by placing a folded piece of filter paper close to the mica sheets. The appearance of a slightly brown colour on the exposed paper surface is an indication of the production of carbon film. A good criterion is that as soon as a difference in colour is noticed on the filter paper the carbon film will be thick enough (~ 10 nm thick). At this point, remove the mica sheets from the evaporator and keep them in a Petri dish.

(iv) In order to obtain good adsorption onto the grids (Section 3.2.3.), it is necessary to wait $24-48$ h after coating before using the carbon film.

3.1.3 *Deposition of the Carbon Film on Cleaned Grids*

The carbon film prepared on the mica sheets is detached from the mica by flotation on a water surface and then deposited on the electron microscope grids by lowering the water level (*Figure 2*). To do this:

(i)   Place a sheet of filter paper (Schleicher and Schüll) on a stainless steel wire net inside a glass beaker (10 cm in diameter and at least 5 cm deep). For ease of visualisation, stand the beaker on a black surface.

(ii) Fill the beaker with distilled water and place the grids (previously cleaned as described in Section 3.1.1) on the filter paper, shiny side up (*Figure 2a*).

(iii) Introduce the carbon-coated mica sheet into the water slowly and progressively at an angle of 45° so that the carbon film is floated at the water surface (*Figure 2a*). The position of the floating carbon film is best visualised under the light of a lamp, placed a short distance away.

(iv) Progressively remove the water while maintaining the carbon film above the grids so that it will cover the surface occupied by the grids (*Figure 2b*).

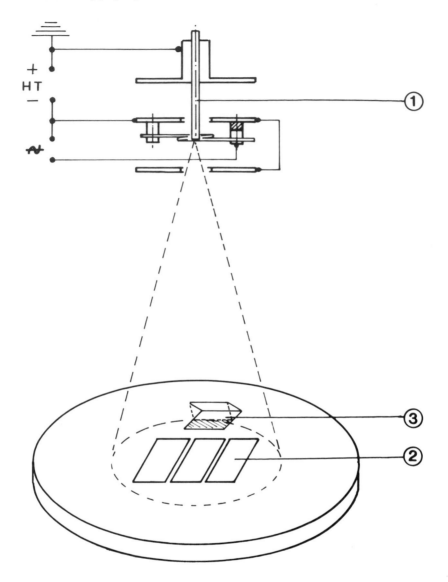

Figure 1. Formation of the carbon film on a freshly cleaved mica sheet. The source of carbon (Edwards electron gun) is represented at the top (**1**). The mica sheets (**2**) are placed at about 20 cm from the source close to a folded piece of filter paper (**3**).

(v)   Recover the filter paper supporting the grids and the carbon film (*Figure 2c*). Place it in a Petri dish and allow it to dry slowly in a clean area.

(vi)   When dry, the grids (now coated with carbon film) can be carefully lifted from the filter paper. This usually leaves a white, round surface distinguishable from the light brown colour of the carbon film.

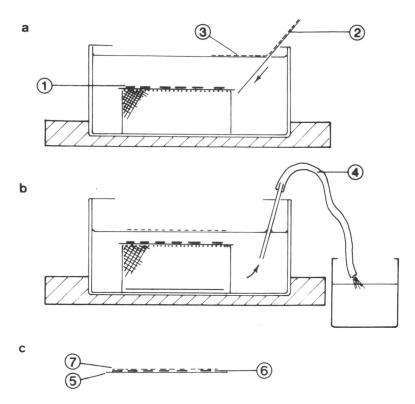

Figure 2. Recovery of the carbon film on the grids. (**a**) The grids (**1**) are placed on a filter paper in a beaker containing water. The mica sheet (**2**) supporting the carbon film is progressively introduced in the water at an angle of 45° to float the carbon film (**3**) onto the water surface. (**b**) The recovery of the carbon film on the electron microscopic grids is achieved by lowering the water level using a siphon (**4**). (**c**) The filter paper (**5**) is recovered bearing the grids (**6**) covered by the carbon film (**7**).

3.2 **Spreading of a Protein Monolayer**

3.2.1 *Solutions for Spreading*

Cytochrome C. In spreading techniques, the slightly basic protein, cytochrome C, is the protein most commonly used. Prepare a stock solution of horse heart cytochrome C (type VI Sigma) in double-distilled water to 1 mg/ml. Because of the great variability in quality and hydration of different batches of commercial cytochrome C, the final concentration should be measured spectrophotometrically and adjusted in double-distilled water to 1 mg/ml using the following relationship (2):

$$E\,^{0.01\%}_{408\,\text{nm}} = 1.25$$

Finally, filter the solution through a disposable filter (0.22 μm pore size; Millipore). It can be stored at 4°C for several months.

Formamide. It is necessary to de-ionise commercially-available formamide in order to avoid aggregation of nucleic acids and irregular spreading of single-stranded molecules. This can be achieved most easily by treatment with a mixed-bed ion-exchange resin (8). Add 5 g of Dowex (AG501, Bio Rad) to 100 ml of formamide and gently mix for 2 h at room temperature. After filtration through filter paper, the solution is stored in 0.5 ml aliquots at −20°C until use.

3.2.2 *Glass Slides used for Spreading*

Glass slides commonly used for light microscopy (7.5 cm x 2.5 cm) are suitable for spreading the protein-nucleic acid complexes. Commercially pre-washed glass slides (Superior) must still be carefully cleaned in order to obtain reproducible and homogeneous films:

(i)   Place the glass slides in a glass slide holder inside a beaker. Add enough chromic acid to completely cover the glass slides and then cover the beaker with aluminium foil.

(ii)   Heat the solution in a fume cupboard and allow it to boil for 20 min.

(iii)   Allow the acid to cool down slightly. Then discard it and rinse the slides extensively with tap water (at least 1 h).

(iv)   Rinse the slides in sterile distilled water.

(v)   The slides can be stored in distilled water for several days until use.

3.2.3 *Spreading Technique*

The protein monolayer can be produced using several different methods such as the single-drop technique (19) or the microdrop diffusion technique (20). However, for DNA-DNA hybrids and DNA-RNA hybrids, the authors prefer to spread a solution containing formamide (the hyperphase) along a clean glass ramp onto distilled water (the hypophase). This technique can be learned easily by any researcher in a few days. The exact procedure is as follows:

(i)   Fill a Teflon trough (10 cm long x 4 cm wide x 1 cm deep) entirely with double-distilled water. Teflon is chosen as a hydrophobic and easily cleaned material.

(ii)   Dip a clean glass slide halfway into the hypophase, making a 20° angle with the water surface (*Figure 3a*).

(iii)   Aspirate the excess water using a Pasteur pipette connected to a water pump, lowering the water level by 3 mm and leaving a 1−2 cm portion of the glass slide wet (*Figure 3b*).

(iv)   Prepare 50 μl of spreading reaction mixture (hyperphase) just before use. This contains:

        25 μl deionised formamide

        2.5 μl 0.2 M EDTA, 2 M Tris-HCl, pH 8.5

        14 μl double-distilled water

        1 μl of internal standards (see Section 4)

        5 μl of hybridisation mixture (see Section 2)

        2.5 μl cytochrome C at 1 mg/ml

(v)   Immediately after mixing, suck this reaction mixture into a 50 μl glass capillary connected to a syringe pump (Razel Scientific Instruments Inc.). The pump speed can be adjusted in order to spread the 50 μl of hyperphase solution in 5 to 10 sec.

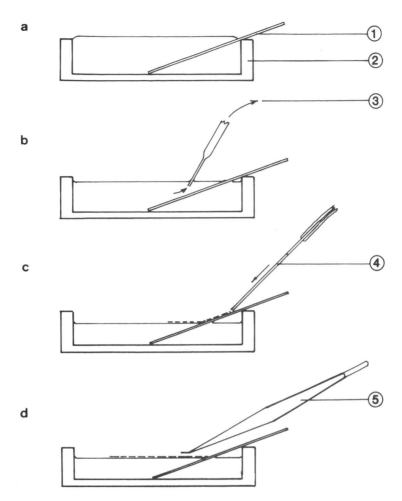

Figure 3. Schematic representation of the spreading of a protein-film. (**a**) A clean glass slide (**1**) is introduced into the Teflon trough (**2**) filled with distilled water. (**b**) Excess water is removed using a Pasteur pipette (**3**) connected to a water pump, thereby exposing 1−2 cm of wet glass slide to the air. (**c**) The spreading is achieved by moving the capillary (**4**) back and forth along the glass slide just above the water surface, releasing the spreading solution at constant slow speed. (**d**) The film is adsorbed onto grids held in forceps (**5**) by touching the surface 1−2 cm from the glass slide.

(vi)   Spread the film by placing the glass capillary very close to the water surface along the glass slide at an angle of 45° with the water (*Figure 3c*). Expel the solution at a constant speed, moving the capillary back and forth along the glass slide. Formation of the protein film and the area of surface occupied can be seen initially by the foaming produced by the salt and formamide solution.

(vii)   Leave the protein film to stabilise for 1 min.

(viii)   Bend one edge of a carbon-coated grid up so that the grid can be held by forceps horizontally to the film present at the air-water interface. The grid needs to just touch the surface so that the protein film is picked up on the grid (*Figure 3d*).

171

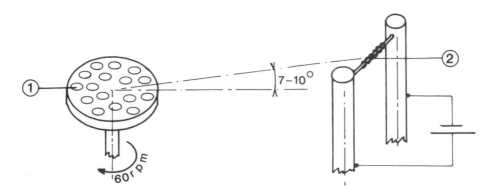

Figure 4. Rotary shadowing of the grids. The grids (**1**) are shadowed under a sharp angle (7–10°) by platinum wire wrapped around a tungsten filament (**2**).

(ix)   Immediately process the grid through steps (iii) and (iv) of the staining procedure (Section 3.3 below). For each spreading, prepare three or four samples successively, touching different regions between 1 and 2 cm in front of the glass slide-water limit.

3.3 Staining Procedure

The contrast of spread nucleic acids in the electron microscope can be increased by staining with a heavy metal before rotary shadowing. The most commonly used stain is uranyl acetate.

(i)   Prepare a stock solution of 0.5% uranyl acetate in 50 mM HCl (6). After dissolution, filter the solution through a disposable filter (0.22 μm; Millipore) and store it at room temperature in a dark bottle covered by aluminium foil to prevent exposure to light. It is stable for 2–3 months.

(ii)   Just before use, prepare the staining solution by mixing 0.5 ml of the uranyl acetate stock solution with 49.5 ml of 90% ethanol in water. This solution must be used within 60 min of preparation and at room temperature.

(iii)   Replace the drop of water present on the surface of the grid [Section 3.2.3, step (ix)] with the staining solution by touching the surface of the staining solution for a few seconds.

(iv)   Dip the grid into a small beaker containing the same staining solution for at least 1 min.

(v)   Next, rinse the grid for 1 min in 90% ethanol, then apply it to a sheet of filter paper to remove the excess solution. Allow it to air dry.

3.4 Rotary Shadowing

To enhance the contrast of spread nucleic acids, the grids may be rotary shadowed with platinum/tungsten (*Figure 4*) before examination by electron microscopy:

(1)   Place the grids flat on a rotary support at a distance of 0.1 mm from and at an angle of 7°–10° to the source. Wrap a piece of platinum wire (10 cm long, 0.1 mm thick) around a 1 mm thick tungsten filament.

(ii)   Carry out the platinum vaporisation under a vacuum of 10^{-6} Torr with the grid support rotating at 60 r.p.m. Increase the voltage gradually until melting of the platinum is observed through dark glasses. Wait 5 sec and then double the voltage for $5-10$ sec. The grids are now ready to be examined by electron microscopy.

4. ELECTRON MICROSCOPIC ANALYSIS

In the authors' laboratory, electron microscope grids are usually examined using a Philips 301 electron microscope fitted with an anti-contamination device. Typically the observation is carried out at 9000- to 20 000-fold magnification with an accelerating voltage of 60 kV and an objective aperture of 20 μm.

The apparent length of any given DNA molecule can vary depending on conditions that affect the surface tension of the protein film. Since the hyperphase contains a high concentration of formamide and the hypophase consists of water, local variation in the organisation of the protein film can occur, introducing local variation in the measured lengths of single- and double-stranded nucleic acids. Therefore, nucleic acid lengths are mostly measured as lengths relative to a single- and/or double-stranded internal standard DNA of known genome length (e.g., SV40, plasmid, phage ϕX174 or fd phage) present in the spreading solution at a concentration such that each hybrid can be standarised by a nearby standard. We routinely co-spread single-stranded ϕX174 DNA and double-stranded ϕX174 RF DNA to provide single- and double-stranded DNA standards. The number of base pairs per unit length is thus determined for double- and single-stranded molecules and verified in each area of the grid. Then a minimum of 30 hybrid molecules are measured. This is necessary to obtain the statistical significance of each measurement. In addition, each molecule must be measured two or three times in order to eliminate human errors in the drawing of each structure, and measurement by two independent researchers is advisable to avoid possible subjective selection of structures. The measured lengths of nucleic acid molecules should vary around a mean with a reproducible standard deviation. Two alternative procedures can be used to determine the lengths:

(i)   Length measurements may be made directly on a TV monitor. The authors' electron microscope is equipped with a TV camera displaying the images on a monitor in front of which is placed a transparent digitising tablet. The digitising tablet (Digistrand Matra Optique SA) is coupled to a minicomputer system which enables the operator to perform on line length measurements without any photographic process. The measurements are stored and processed by interactive programs. The whole system was developed by the Laboratoire d'Applications Electroniques. The programs, written in Fortran, are available from the authors on request.

(ii)   Using a grid in which the nucleic acid molecules are well spread and not too confluent, take a series of micrographs of the sample and of the internal standards (double- and single-stranded) without changing the alignment of the electron microscope. Prepare enlarged prints from the photographic negatives and use these to trace and measure the length of the molecules using a map measurer with 1 mm resolution. Alternatively, the same pictures can be analysed using a digitising tablet and the computer programs described above.

5. EXAMPLES

5.1 **Heteroduplex Analysis**

Heteroduplex formation has been extensively used to characterise rearrangements in DNA sequences such as deletions, insertions and duplications. For example, we have

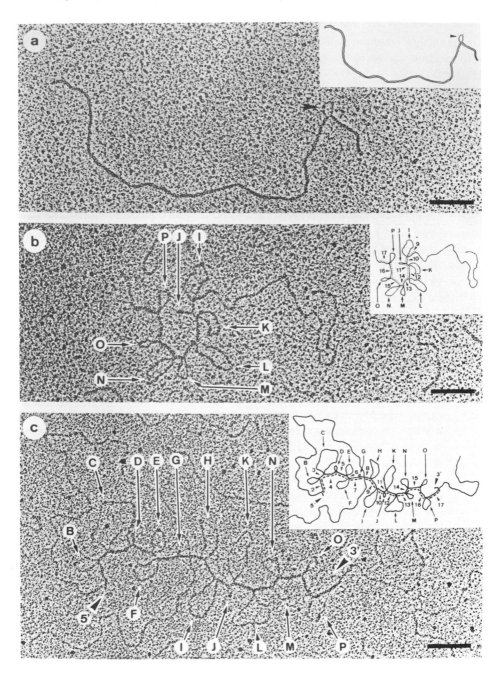

visualised a deletion present in a natural variant, Formosa, of *Drosophila melanogaster* in the *sgs3* gene by hybridisation with the corresponding wild-type sequences obtained from the Oregon R strain (*Figure 5a*). Another example is the characterisation of the arrangement of conalbumin coding sequences along the corresponding genomic fragment by hybridisation between genomic sequences and homologous cDNA. Clear single-stranded loops (introns) are branched out from a thicker double-stranded segment corresponding to the coding sequence. The position of the loops and their lengths provide information about the intron-exon organisation of the conalbumin gene (*Figures 5b* and *6b*).

5.2 DNA-RNA Hybrids

DNA-RNA hybridisation may be used to analyse the molecular structure and organisation of genes. For example, *Figures 5c* and *6a* show R-hybrid molecules between a genomic fragment of the conalbumin gene and pheasant conalbumin mRNA. The orientation $(5' \rightarrow 3')$ of a gene may be deduced by the presence of a short single-stranded tail corresponding to the poly(A) chain at the 3' end (as in *Figure 5c*) or by comparing the structures formed between single-stranded genomic DNA and mRNA (DNA-RNA hybrid) and genomic DNA and cDNA (DNA-cDNA heteroduplex) using the same genomic DNA fragment for the split gene (the 5' end of the gene often being absent in the cDNA sequences).

6. LIMITATIONS TO HYBRID ANALYSIS

Electron microscopic characterisation of nucleic acid hybrids provides information on the homology between the two participating molecules. Essentially two factors limit the analysis:

(i)   A hybrid region must exceed a minimum length in order to be recognised. The stability of this minimum length depends to a large extent on the hybridisation conditions.

(ii) If the homologous regions are not fully complementary, this will affect the minimum hybrid length that can be visualised.

Increasing or decreasing the stringency of the hybridisation conditions will decrease or increase the minimum observable hybrid lengths, respectively. These limitations are discussed in more detail below, together with the effects of the G+C content of the nucleic acids being examined and the presence of intramolecular repeat sequences.

Figure 5. Electron microscopy of hybrid molecules. (**a**) Hybrid molecule between the *sgs3* genes from Formosa and Oregon R strains of *Drosophila* (from ref. 21). A deletion of about 300 bp is visualised by a single-stranded loop indicated by the arrow. (**b**) Heteroduplex molecule between pheasant conalbumin cDNA fragment (1.3 kb) and a conalbumin genomic fragment. The coding sequences are visualised as double-stranded (thicker filaments numbered 9−17 in the inset diagram) and the intervening sequences as single-stranded loops (letters I−P). The pheasant conalbumin genomic and cDNA were cloned by Daniel Dupret in our laboratory (manuscript in preparation). (**c**) DNA-RNA hybrid molecule between a pheasant genomic fragment of conalbumin gene and conalbumin mRNA. As in (**b**), the coding sequences show double-stranded structures (exons 2−17 in the inset diagram), while the intervening sequences appear as branched loops (introns B−P). A short free single-stranded tail corresponding to the poly(A) chain is observed at the 3' end of the gene (marked 3'). In each part, the inset shows a schematic representation of the nucleic acid structures seen in the electron microscope with the continuous line representing the genomic sequences, the dotted line cDNA and the dashed line mRNA sequences. The bar represents 0.1 μm.

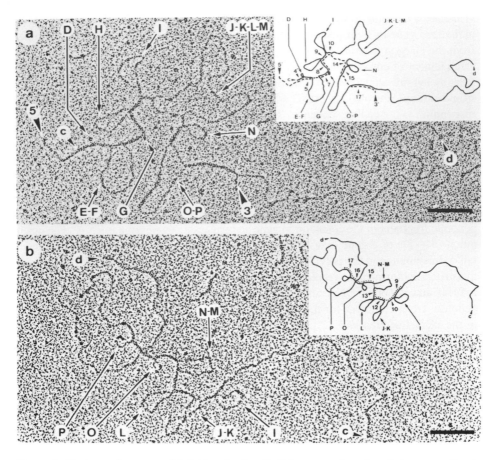

Figure 6. Electron microscopy of hybrid molecules. Partial sequence homology between conalbumin sequences from chicken and pheasant. (**a**) DNA-RNA hybrid between pheasant conalbumin mRNA and a chicken conalbumin genomic fragment. As for *Figure 5b* and *c*, the introns show up as branched single-stranded loops (D-H-I-N). Three regions marked E-F, J-K-L-M and O-P present a different structure: the two segments of the single-stranded loops have different lengths. The coding sequence (shortest side of the loop) contained in the mRNA molecule did not hybridise with the corresponding sequences in the genomic fragment. Thus the longest part of the loop contains both the non-homologous coding sequences and the corresponding introns. A short tail of poly(A) is visualised at the 3' end of the molecule. (**b**) DNA-DNA hybrid between pheasant conalbumin cDNA and the chicken conalbumin gene. As represented in the insert, exons 11 and 14 did not hybridise leaving loops marked N-M and J-K. The continuous line represents the genomic sequences, the dotted line cDNA and the dashed line mRNA sequences. The bar represents 0.1 μm.

6.1 The Minimum Stable Hybrid Length

The shortest DNA-DNA or DNA-RNA hybrid we have visualised is a 30 ± 10 bp region of exon 11 of the conalbumin gene. This region was found in a larger complex containing several hybrid and loop structures which was certainly responsible, at least in part, for the stabilisation of such a short hybrid. Therefore, very short regions of homology are more easily recognised in long DNA fragments when other complementary segments exist along the same molecules. The visualisation of hybrids about 40 bp long have also been described in other systems, for example, the J segments of λ and *x* immunoglobulin genes (22) and the leader sequence of adenovirus-2 late mRNA (23).

6.2 **Degree of Homology**

Homology between two DNA sequences has been studied by heteroduplex formation (e.g., between two phages T7 and T3; see ref. 24) and RNA-DNA homologies have been determined by R-loop mapping (e.g., detection of homologies between J sequences using purified mRNA of embryonic mouse immunoglobulin genes; see refs. 25, 26). In these cases, the degree of sequence homology was related to the percentage of mismatch between the two nucleotide sequences. As mentioned above, heteroduplex formation is dependent on the stringency of the hybridisation conditions. Under the hybridisation conditions described in this chapter, DNA-DNA heteroduplexes are formed when the sequence homology exceeds 65%, that is, less than 35% mismatch. In contrast, DNA-RNA hybrid sequences must have less than 20% mismatch to be observed as perfect hetero-hybrids. Under less stringent conditions (e.g., lowering the temperature by a few degrees or the formamide concentration by a few percent), heteroduplexes and hetero-hybrids can be formed with a lower percentage of homology between the two sequences.

An example of the determination of homology is shown in *Figure 6* where the homology between the conalbumin genes of chicken and pheasant was studied both by DNA-DNA and RNA-DNA hybridisation (27). Heterohybrids were formed between the chicken genomic conalbumin gene and the pheasant conalbumin mRNA (*Figure 6a*) or cDNA (*Figure 6b*). A strong homology in the conalbumin sequence and gene arrangement was detected between the two species. In both cases (DNA-DNA heteroduplexes and DNA-RNA hybrids), exon 11 (30 ± 10 bp) is the only hybrid sequence absent, presumably because the small size of homologous sequence between the two species leads to an unstable hybrid in this exonic region.

6.3 **G+C Content**

The stability of a hybrid increases with increasing G+C content of the complementary regions (15). Generally the relative G+C richness of the exon *versus* intron regions will favour the stability of the exonic hybrid sequences of eukaryotic genes.

6.4 **Intramolecular Homologous Sequences**

The existence of direct or inverted repeat sequences in the same molecule can complicate the analysis of nucleic acids by electron microscopy since these give rise to intramolecular hybrid regions. Intramolecular repeated sequences can be detected using the heteroduplex technique described above.

7. ACKNOWLEDGEMENTS

The authors are indebted to D.Dupret for a gift of pheasant conalbumin cDNA and genomic clones and thank M.Bourouis and G.Richards for *Drosophila* clones. We would also like to thank J.C.Homo for the diagrams, B.Boulay for photographic expertise and E.Badzinski and C.Kutschis for typing the manuscript. This work was supported by grants from the Ministère de l'Industrie et de la Recherche (contract no. 83.V.0626) and CNRS (ATP no. 6182).

8. REFERENCES

1. Kleinschmidt,A.K. and Zahn,R.K.S. (1959) *Naturforschung,* **146**, 770.
2. Everson,D.P. (1977) *Methods Virol.,* **6**, 219.
3. Ferguson,J. and Davis,R.W. (1978) in *Advanced Techniques in Biological Electron Microscopy II,* Springer-Verlag, Berlin/Heidelberg, p. 123.
4. Fisher,M.W. and Williams,R.C. (1979) *Annu. Rev. Biochem.,* **48**, 649.
5. Brack,C. (1981) *Crit. Rev. Biochem.,* **10**, 113.
6. Davis,R.W., Simon,M. and Davidson,N. (1971) in *Methods in Enzymology,* Vol. **21**, Grossman,L. and Moldave,K. (eds.), Academic Press Inc., NY, p. 413.
7. Garapin,A.C., Cami,B., Roskam,W., Kourilsky,P., LePennec,J.P., Perrin,F., Gerlinger,P., Cochet,M. and Chambon,P. (1978) *Cell,* **14**, 629.
8. Westmoreland,B.W., Szybalski,W. and Ris,H. (1969) *Science (Wash.),* **163**, 1343.
9. Wellauer,P.K. and Dawid,I.B. (1977) *Cell,* **10**, 193.
10. McConaughy,B.L., Laird,C.D. and McCarthy,B.J. (1969) *Biochemistry (Wash.),* **8**, 3289.
11. Kleinschmidt,A.K. (1968) in *Methods in Enzymology,* Vol. **12B**, Grossman,L. and Moldave,K. (eds.), Academic Press Inc., NY, p. 361.
12. Thomas,M., White,R.L. and Davis,R.W. (1976) *Proc. Natl. Acad. Sci. USA,* **73**, 2294.
13. White,R.L. and Hogness,D.S. (1977) *Cell,* **10**, 177.
14. Holmes,D.S., Cohn,R.H., Kedes,L.M. and Davidson,N. (1977) *Biochemistry (Wash.),* **16**, 1504.
15. Casey,J. and Davidson,N. (1977) *Nucleic Acids Res.,* **4**, 1539.
16. Garapin,A.C., LePennec,J.P., Roskam,W., Perrin,F., Cami,B., Krust,A., Breathnach,R., Chambon,P. and Kourilsky,P. (1978) *Nature,* **273**, 349.
17. Baumester,W. and Hahn,M. (1978) *Principles and Techniques of Electron Microscopy,* Van Nostrand Reinhold, New York.
18. Fukami,A. and Adachi,K. (1965) *J. Electron Microsc.,* **14**, 112.
19. Inman,R.B. and Schnös,M. (1970) *J. Mol. Biol.,* **49**, 93.
20. Lang,D. and Mitani,M. (1970) *Biopolymers,* **9**, 373.
21. Richards,G., Cassab,A., Bourouis,M., Jarry,B. and Dissous,C. (1983) *EMBO J.,* **2**, 2137.
22. Sakano,H., Rogers,J.H., Hüppi,K., Brack,C., Trennacker,A., Mahi,R., Wall,R. and Tonegawa,S. (1978) *Nature,* **277**, 627.
23. Berget,S.M., Moore,C. and Sharp,P.A. (1977) *Proc. Natl. Acad. Sci. USA,* **74**, 3171.
24. Davis,R.W. and Hyman,R.W. (1971) *J. Mol. Biol.,* **62**, 287.
25. Sakano,H., Hüppi,K., Heinrich,G. and Tonegawa,S. (1979) *Nature,* **280**, 288.
26. Max,E.E., Seidman,J.G. and Leder,P. (1980) *Proc. Natl. Acad. Sci. USA,* **77**, 2138.
27. Maroteaux,L., Heilig,R., Dupret,D. and Mandel,J.L. (1983) *Nucleic Acids Res.,* **11**, 1227.

CHAPTER 8

In Situ Hybridisation

MARY LOU PARDUE

1. INTRODUCTION

Hybridisation of a nucleic acid probe to nucleic acids within cytological preparations permits a high degree of spatial localisation of sequences complementary to that probe. This localisation can be useful in a number of ways for answering biological questions. For example, *in situ* hybridisation to the DNA of condensed chromosomes can be used to map the sites of particular sequences (*Figures 1* and *2*). Hybridisation to the DNA of interphase nuclei can be used to study the functional organisation of specific sequences within the diffuse chromatin that characterises this stage of the cell cycle (*Figure 3*). *In situ* hybridisation to cellular RNA (*Figure 4*) allows a very precise analysis of the tissue distribution of any RNA species of interest. In addition, *in situ* hybridisation makes it possible to study the RNA of individual cells unaffected by the RNA of other cells in the tissue. Thus it should be possible to use the technique to detect RNAs that are present in only a very small subset of cells. Such RNAs might never be detected

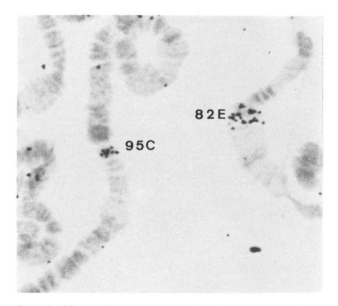

Figure 1. Autoradiograph of *Drosophila mauritainia* polytene chromosomes showing two sites coding for the small nuclear RNA, UI, in regions 95C and 82E. A third site (21E) does not appear in this photograph. The chromosomes were hybridised with plasmid pBR322 carrying a 90-bp fragment of DNA encoding nucleo-tides 50 − 139 of UI RNA (cloned by Dr S. Mount). The DNA had been nick-translated with [³H]TTP. Exposure time 25 days; magnification x1300.

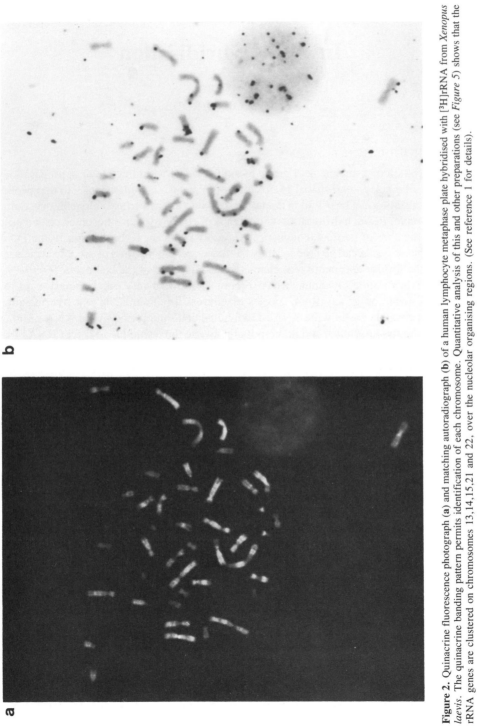

Figure 2. Quinacrine fluorescence photograph (**a**) and matching autoradiograph (**b**) of a human lymphocyte metaphase plate hybridised with [³H]rRNA from *Xenopus laevis*. The quinacrine banding pattern permits identification of each chromosome. Quantitative analysis of this and other preparations (see *Figure 5*) shows that the rRNA genes are clustered on chromosomes 13,14,15,21 and 22, over the nucleolar organising regions. (See reference 1 for details).

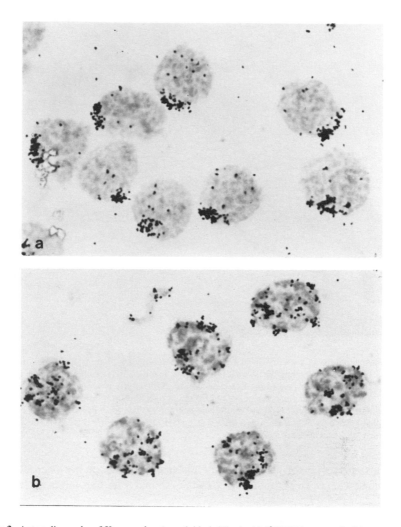

Figure 3. Autoradiographs of *Xenopus laevis* nuclei hybridised with [³H]RNA transcribed from genes coding for 5S RNA. Clusters of these genes are located at the ends of the long arms of most, if not all, of the chromosomes. Hybridisation allows localisation of these chromosome ends even in diffuse chromatin. (a) Zygotene nuclei from the testis. In this cell type the chromosome ends are tightly clustered on one side of the nucleus. (b) Spermatogonia from the testis. In this cell type the chromosome ends are much more dispersed.

in RNA extracted from a whole tissue because of dilution by other RNA species from the vast majority of the cells which do not contain the RNA of interest. Thus, *in situ* hybridisation may be useful not only to show where an RNA is localised but, in some cases, it may be the best way to show that the RNA exists at all.

The two types of *in situ* hybridisation that are in general use, hybridisation to nuclear DNA and hybridisation to cellular RNA, are conceptually quite similar but differ significantly in technical details. For this reason the two techniques will be discussed separately in this chapter. Most *in situ* hybridisation is done using preparations that are analysed with the light microscope. Recently a technique for hybridisation to

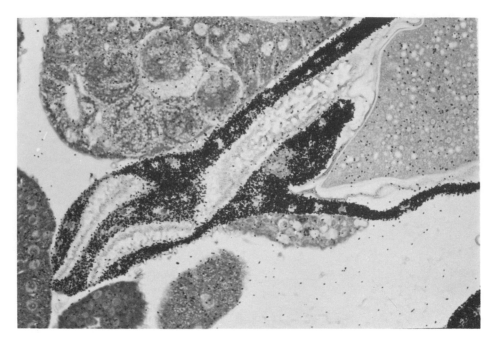

Figure 4. Autoradiograph of a section from a *Drosophila melanogaster* ovary hybridised with [^{3}H]cRNA complementary to mRNA encoding the chorion protein S15. The protein is synthesised by the layer of follicle cells which covers the surface of the late stage oocyte. S15 mRNA is not detected in the other cells of the ovary. The autoradiograph shows heavy hybridisation over the cytoplasm of the follicle cells. The small unlabelled spots in this cell layer are the follicle cell nuclei. The [^{3}H]cRNA was transcribed from a recombinant M13 phage by *E. coli* RNA polymerase using [^{3}H]UTP. Exposure time 25 days, magnification x560. (Courtesy of J.G. Gall and S. Parks).

chromosomes prepared for analysis by the electron microscope has been developed. That technique will not be discussed here; readers should refer to reference 2 for details.

2. EQUIPMENT

2.1 Subbed Slides

The surface of these slides gives better retention of cytological preparations. Microscope slides are washed either in acid or in detergent and thoroughly rinsed in water. They are put into a final rinse of distilled water and then dipped into subbing solution. This is an aqueous solution of 0.1% gelatin and 0.01% chrome alum [chromium potassium sulfate; $CrK(SO_4)_2 \cdot 12 H_2O$]. The gelatin is first dissolved in hot water ($\sim 65\,^\circ C$) and the chrome alum is added after the solution has cooled. Subbing solution can be stored at 4°C for long periods. After dipping in subbing solution, the slides are allowed to dry for several hours before being used. These slides can be stored for long periods in a dry, dust-free container.

2.2 Siliconised Coverslips

Siliconisation prevents adherence of cytological material to the coverslips. The coverslips can be treated using any of the procedures used to siliconise laboratory glassware. For example, immerse the coverslips briefly in a 2% solution of Functional Silane (PCR Research Chemicals) and allow them to dry overnight at room temperature. The siliconised coverslips are stored dry. Immediately before use, each coverslip is wiped free of dust using microscope lens tissue. If desired, the coverslips can be rinsed with 95% ethanol before wiping with the lens tissue. Note that the Functional Silane solution tends to degrade upon exposure to air so that after the stock bottle has been opened the remainder should be stored under nitrogen.

2.3 Polylysine-coated Slides

Polylysine-coated slides are used in only certain protocols for cytological preparations. They can be prepared as follows. Soak glass slides in detergent for at least 2 hours. Then wash in running tap water (2 h), two changes of distilled water (5 min each) and two changes of 95% ethanol (5 min each). Air-dry in a dust-free environment. Prepare fresh poly-L-lysine hydrobromide (150 000 − 300 000 mol. wt. fraction) at 0.5 mg/ml in distilled water (19,20). Dip the slides in the polylysine solution. Air-dry. Store in a dust-free environment at 4°C.

2.4 Moist Chambers

Moist chambers are used to prevent evaporation when the cytological preparation is being incubated in a small amount of liquid. Small plastic sandwich boxes are ideal. Each box should have a tight-fitting lid so that it can maintain a moist atmosphere. The box should be made of plastic since moisture condensing on a plastic lid will remain in small droplets; moisture condensing on glass lids tends to coalesce into large drops which may fall onto the cytological preparation. The bottom of the chamber is covered with a thin layer of the incubation medium. The slides are supported above the liquid on small objects such as plastic tube caps. A drop of incubation medium (e.g., hybridisation mixture or an enzyme-containing solution) is placed over the cytological preparation and covered with a coverslip to spread the liquid evenly. It is important that the liquid in the bottom of the moist chamber has the same salt concentration as the solution under the coverslip to prevent distillation and subsequent changes in concentration in the solution covering the cytological preparation.

3. NUCLEIC ACID PROBES FOR IN SITU HYBRIDISATION

3.1 Choice of Label for Probes

Tritium is by far the best radioisotope for *in situ* hybridisation because of the extremely low energy of the beta particle that is emitted (0.0181 MeV). These beta particles travel less that 1 μm through the autoradiographic emulsion, ensuring that the silver grains in the autoradiograph remain closely associated with the site of the radioactive molecule. [125I]Iodine has also been used for *in situ* hybridisation but the radiation emitted by 125I is significantly more energetic than that emitted by 3H and therefore 125I gives less precise cytological localisation. 125I also gives a much higher background.

Techniques have also been developed for the detection of nucleic acids by non-autoradiographic means. The most successful of these uses biotin-substituted nucleotides to label nucleic acid probes. The probe is then detected by biotin-specific antibodies conjugated to fluorescent or enzymatic reagents (3). This technique has been used successfully for both light microscope and electron microscope studies (2,3). The method has the advantage of rapid analysis since it is not necessary to wait for an autoradiograph to expose. However, at the present time, detection by non-autoradiographic means is significantly less sensitive than autoradiography and cannot be used except where a large amount of hybrid is formed. In addition, non-autoradiographic detection is not as easily quantitated as is autoradiography and therefore does not yield all of the information that can be obtained from *in situ* hybridisation with radioactive probes.

3.2 **Choice of Probe**

Recombinant DNA technology now provides the opportunity to obtain either DNA or RNA probes of any desired sequence. Furthermore one can choose between single-stranded and double-stranded probes. Which should be used for an *in situ* hybridisation experiment? There are various considerations. RNA probes produce less background than DNA probes. RNase treatment of the preparation after the hybridisation step efficiently removes non-specifically bound RNA. Non-specifically bound DNA probe is best removed by washes at a temperature slightly less than the hybridisation temperature. The single strand-specific nuclease S1 is *not* effective in removing unhybridised DNA from cytological preparations. Double-stranded probes, if randomly sheared, may form networks on the cytological hybrid and so increase the hybridisation signal. On the other hand, double-stranded probes can also anneal in solution and thus reduce the concentration of probe available for reaction with the cytological preparation.

For hybridisation to DNA in cytological preparations, double- and single-stranded probes, both DNA and RNA, have been used successfully. There has been little attempt to study systematically the relative efficiency of these different probes. Both double- and single-stranded probes can also be used for hybridisation to RNA in cytological preparations. However, for RNA studies, single-stranded probes give increased assurance that the hybrid formed is that desired, provided it is known which DNA strand is copied into RNA. A probe complementary to the transcribed strand should not hybridise with the cellular RNA and therefore makes a useful control. For hybridisation to RNA in cytological preparations it is generally agreed that very short probes (50 − 150 nt) yield the most efficient hybridisation. DNA probes can be reduced in size by controlled nicking with DNase I. Small RNA probes can be prepared by limited alkaline hydrolysis. The RNA can be hydrolysed in 40 mM $NaHCO_3$, 60 mM Na_2CO_3 (pH 10.2) at 60°C. Hydrolysis time is calculated from the formula

$$t = \frac{L_0 - L_f}{k \, L_0 L_f}$$

where t is the time in min and L_0 and L_f are the initial and final fragment lengths in kb (4). The rate constant for hydrolysis, k, is approximately 0.11 kb/min. After hydrolysis, neutralise the samples by adding sodium acetate (pH 6.0) to 0.1 M final concentration and glacial acetic acid to 0.5% (v/v). Then add 2.5 volumes of ethanol and leave the mixture at −20°C for at least 1 h to precipitate the probe. The final length of the probe can be checked by electrophoresis in 10% polyacrylamide gels in 8 M urea at room temperature (ref. 5).

3.3 **Techniques for Labelling Probes**

Detailed protocols for the many techniques available for labelling nucleic acid probes are given in Chapter 2. The following sections describe variations of these techniques used in the author's laboratory specifically to label probes for *in situ* hybridisation.

3.3.1 *In Vitro Transcription by E. coli RNA Polymerase*

Both double-stranded DNA and single-stranded DNA (such as M13 clones) can be more-or-less randomly transcribed by *E. coli* RNA polymerase (6). A typical labelling protocol is given in *Table 1*.

3.3.2 *In Vitro Transcription with SP6 RNA Polymerase*

An efficient way to obtain a single-stranded RNA probe for a known DNA sequence

Table 1. Preparation of Labelled Probes using *E. coli* RNA Polymerase.

1.	To a microcentrifuge tube, add: 50 μCi [^{3}H]UTP (50 Ci/mmol) 5 μl of a mixture of 1 mM ATP, CTP and GTP Dry down under vacuum.
2.	Add to the dried ribonucleotides: 5 x salts solution [a]　　　　　18.4 μl 0.125 M MnCl$_2$　　　　　　　1.6 μl 0.2% 2-mercaptoethanol　　　20 μl Cloned DNA　　　　　　　　$1-2$ μg *E. coli* RNA polymerase　　　$1-2$ units Water to 100 μl final volume
3.	Incubate at 37°C for $1-2$ h.
4.	Add 200 μl of 10 mM Tris-HCl, pH 7.9, and 10 μl DNase (1 mg/ml; RNase-free). Leave at room temperature for $10-15$ min.
5.	Spot 1 μl aliquots onto two nitrocellulose filters. Dry one filter. Rinse the other in ice-cold 5% TCA for 5 min, wash in 70% ethanol and dry. Count both filters. The proportion of the radioactivity that is TCA-insoluble will vary but should be at least 10%. The incorporation is usually much higher.
6.	Add 10 μl 0.2 M EDTA, 20 μg *E. coli* tRNA (20 mg/ml), 25 μl 10% SDS, 300 μl phenol-chloroform-isoamylalcohol (25:24:1 by vol). Mix by vortexing (1 min).
7.	Centrifuge and remove the top (aqueous) layer into a clean microcentrifuge tube.
8.	Add an equal volume of chloroform:isoamyl alcohol (24:1, v/v).
9.	Centrifuge for 5 min at 10 000 *g*. Remove the chloroform (bottom layer).
10.	Repeat the chloroform extraction (step 9).
11.	Add 600 μl ethanol and leave at -20°C for at least 1 h.
12.	Centrifuge for 25 min at 10 000 *g*. Discard the supernatant.
13.	Dissolve the RNA pellet in distilled water that has been treated with diethylpyrocarbonate (DEPC)[b].
14.	Bring to 0.3 M ammonium acetate. Add 2.5 volumes of ethanol. Leave at -20°C for at least 1 h. Repeat steps 12 and 13.
15.	Repeat step 14.
16.	Reduce the overall size of the RNA molecules to $50-150$ nucleotides by limited alkaline hydrolysis as described in Section 3.2.
17.	Recover the RNA by ethanol precipitation (steps 11 and 12).
18.	Dissolve the RNA in DEPC-treated water using only 75% of the volume required for the hybridisation reaction[c].

[a]5x Salt solution is prepared by mixing 4 ml of 1 M Tris-HCl (ph 7.9), 19 ml of 1 M KCl, 0.6 ml of 1 M MgCl$_2$, 0.9 ml of 10 mM EDTA.
[b]Distilled water is treated with DEPC to inactivate any contaminating RNase. The procedure is to add DEPC to 0.1% final concentration and then leave at room temperature for 12 h. Residual DEPC is then destroyed by autoclaving the solution for 15 min.
[c]If the hybridisation buffer is to contain formamide, the RNA should be dissolved in a correspondingly smaller volume of water.

185

Table 2. Preparation of Labelled Probes using SP6 Polymerase[a].

1.	Dry down 50 μCi [^{3}H]UTP (50 Ci/mmol) in a disposable plastic microcentrifuge tube.
2.	At room temperature, add:

5 x transcription buffer[b]	4 μl
1 mM each of ATP, CTP and GTP[c]	4 μl
0.20 M DTT[c]	1 μl
Bovine serum albumin (sterile; 2 mg/ml)[c]	1 μl
RNasin (Promega Biotech)	20 units
Cloned DNA[c,d]	1 − 3 μg
SP6 polymerase	3 units
DEPC-treated water to 20 μl final volume.	

3.	Incubate for 1 h at 40°C.
4.	Add 75 μl DEPC-treated water, 40 units DNase (RNase-free; Worthington) and 75 units RNasin. Incubate for 10 min at 37°C.
5.	Estimate the incorporation of radioactivity by determining the TCA-precipitable counts as in *Table 1*, step 5.
6.	Extract the mixture with an equal volume of phenol-chloroform-isoamyl alcohol (25:24:1 by vol.) Mix by vortexing for 1 min. Separate the phases by centrifugation.
7.	Extract the upper (aqueous phase) twice with chloroform.
8.	Add 50 μg yeast tRNA as carrier and then add stock ammonium acetate to 0.7 M final concentration.
9.	Add 2.5 volumes of ethanol and recover the RNA as in *Table 1*, steps 11 and 12.
10.	Redissolve the RNA in DEPC-treated water and repeat the precipitation (steps 8 and 9) twice more but without adding more yeast tRNA carrier.
11.	Finally, dissolve the RNA pellet in DEPC-treated water. Reduce the size of the RNA to 50 − 150 nucleotides by limited alkaline hydrolysis as described in Section 3.2.

[a]An alternative procedure for labelling probes using SP6 polymerase is given in Chapter 2, Section 4.2.2.
[b]The composition of 5x transcription buffer is 0.2 M Tris-HCl (pH 7.5), 30 mM $MgCl_2$, 10 mM spermidine. This buffer is autoclaved and then stored at −20°C.
[c]These solutions are made using DEPC-treated water (see *Table 1,* footnote b).
[d]Circular DNA should be dialysed against 1 mM EDTA, 10 mM Tris-HCl, pH 7.5, to remove salts and then linearised using an appropriate restriction enzyme.

Table 3. Preparation of Labelled Probes by Nick Translation[a].

1.	Dry down 150 pmoles of [^{3}H]TTP and 150 pmoles of [^{3}H]dATP in a plastic microcentrifuge tube.
2.	Add:

0.5 mM dGTP	1 μl
0.5 mM dCTP	1 μl
10 x salts solution[b]	1 μl
1% 2-mercaptoethanol	1 μl
Cloned DNA	0.1 μg
DNase I[c]	1 μl
E. coli DNA polymerase I (10 unit/μl) 1 μl	
Water to 10 μl final volume	

3.	Incubate the mixture at 14°C for 1 to 2 h.
4.	Add 90 μg of carrier DNA[d] and adjust the total volume to 100 μl with water.
5.	Add 3 μl of 0.1 M spermine. Mix and leave on ice for 15 min.
6.	Centrifuge at 10 000 *g* for 10 min. Discard the supernatant.
7.	Resuspend the pellet in 75% ethanol containing 0.3 M sodium acetate and 10 mM magnesium acetate. Vortex. Leave on ice for 1 h, mixing frequently.
8.	Centrifuge for 10 min (10 000 *g*) to recover the DNA.
9.	Resuspend the labelled double-stranded DNA probe in water.

[a]An alternative protocol for preparation of labelled DNA probes by nick translation is given in Chapter 2, Section 4.1.2.
[b]The composition of 10 x salts solution for nick translation is 0.5 M Tris-HCl (pH 7.8), 50 mM $MgCl_2$, 0.5 mg/ml BSA (nucleic acid grade).
[c]Commercial DNase I (1 mg/ml stock) is stored in aliquots at −20°C. Before use, the DNase is diluted by a factor of 10^5 in water.
[d]Use *E. coli* DNA sheared by sonication and denatured by boiling for 10 min.

is to clone the sequence adjacent to a promoter of the phage SP6 (7,8). The SP6 RNA polymerase will only transcribe DNA downstream from this promoter. The end-point of the transcription can be set by cleaving the DNA template with an appropriate restriction enzyme. When preparing a probe for hybridisation to cellular RNA, the promoter should be placed on the non-coding strand of DNA, that is, the strand not transcribed *in vivo*. SP6 polymerase and cloning vectors are commercially available. A typical protocol for labelling probes using SP6 RNA polymerase is given in *Table 2*.

3.3.3 *Nick Translation by E. coli DNA Polymerase I*

Radioactive nucleotides can be introduced into double-stranded DNA by nick-translation with *E. coli* DNA polymerase I. Starting at a nick in the DNA, the enzyme removes nucleotides from the 5′ side of the nick while adding nucleotides to the 3′ side. If this replacement synthesis is carried on in the presence of radioactive precursors, the DNA can be labelled to a high specific radioactivity (9,10). The amount of replacement synthesis is controlled, to a large extent, by the number of nicks which provide entry for the polymerase. The number of nicks depends on the amount of DNase I present in the reaction (or nicks introduced during DNA isolation). A typical protocol is given in *Table 3*.

3.3.4 *Notes on the Specific Radioactivity of Probes*

The specific radioactivity of probes synthesised by RNA polymerases is limited only by the specific activity of the nucleotide precursors. For hybridisation to clustered repeated DNA, to polytene chromosomes, or to an abundant RNA, a probe labelled with only [^{3}H]UTP (50 Ci/mmol) is sufficient. If the target of hybridisation is smaller, it is possible to increase the specific radioactivity of the probe by replacing one or more of the other unlabelled nucleoside triphosphates with additional ^{3}H-labelled nucleoside triphosphates.

The specific radioactivity of DNA labelled by nick translation depends both on the nucleotide precursors and on the extent of replacement of unlabelled DNA. The typical nick-translation reaction described in *Table 3* gives probes labelled to about 3×10^7 c.p.m./μg using [^{3}H]TTP (80 Ci/mmol) and [^{3}H]dATP (14 Ci/mmol). This allows detection of sequences in polytene chromosomes corresponding to a typical phage λ clone in 1−3 days.

The amount of hybrid detected depends both on the specific activity of the probe and on how many of the target sequences have bound the probe. Therefore, it is useful to use the probe at a concentration that will nearly saturate the target DNA. Higher concentrations of probe will only contribute to the background without improving the signal. For hybridisation to polytene chromosomes and to other abundant sequences, the probe is usually used at well below saturating concentrations for reasons of economy. With such large targets, hybrids can be detected after only a few days of autoradiographic exposure even if the DNA is not completely saturated.

4. HYBRIDISATION TO DNA IN CYTOLOGICAL PREPARATIONS

4.1 Cytological Procedures

It is not possible to design a single procedure that makes optimal preparations from all types of cells and from all organisms. Three commonly used cytological procedures are given below. The preparations described are not the only ones that will give good results after hybridisation. If some other procedure yields better cytological preparations of the cells to be studied, it may be better to use that procedure for *in situ* hybridisation of those cells. The procedure adopted should give well spread and very flat preparations since such preparations have both the best morphology and give the highest hybridisation signals. Some fixatives, such as those containing formaldehyde, appear to interfere with the denaturation of DNA and should be avoided. Excessive acid treatments may depurinate the DNA and so reduce the level of hybridisation. Storage of preparations in ethanol may lead to loss of DNA and reduced hybridisation.

4.1.1 *Drosophila Polytene Chromosomes*

(i)   Dissect the larvae in insect Ringer's solution (7.5 g NaCl, 0.35 g KCl, and 0.21 g $CaCl_2$ per litre of water).

(ii)   Transfer the glands to a small drop of 45% acetic acid (the fixative) on a siliconised coverslip (18 mm²). (A small coverslip is used so that the chromosomes will stay in a restricted area). It is important that no Ringer's solution is carried along with the glands since dilution of the acetic acid leads to poor morphology.

(iii)   After the glands have fixed for $2-5$ min, lower a subbed slide (see Section 2) onto the coverslip. Turn the slide over and tap lightly on the coverslip with a pointed object, such as the tip of a ball point pen. Tap directly over the glands. The tapping may move the coverslip sideways slightly. This step is intended to force the chromosomes out of the cells and to spread the chromosome arms. Check the preparation using a phase-contrast microscope.

(iv)   When the chromosomes have been well spread, place the slide on a paper towel with its coverslip side down. Press very hard with the thumb directly over the coverslip. At this point it is important that the coverslip does not slip sideways. The thumb pressure will not improve the spreading of the chromosomes but it will flatten them and thus help with the retention of good morphology.

(v)   Immediately after the thumb press, place the slide on a very flat piece of dry ice with the coverslip side up. Leave on the dry ice until the preparation is frozen to the slide $(2-5$ min).

(vi)   Flip off the coverslip by sliding a razor blade under one corner. Immediately plunge the slide into 95% ethanol at room temperature. Leave for 10 min.

(vii)   Put the slide through two more washes in 95% ethanol, 10 min each.

(viii) Allow the slide to air dry. Slides can then be stored dry for very long periods of time. Preparations can be examined dry using a phase-contrast microscope. The best chromosomes are flat and grey with no refractivity. The banding pattern should be clearly recognisable.

4.1.2 *Human Metaphase Chromosomes*

(i)   Prepare metaphase and prometaphase chromosome spreads from aminopterin-synchronised peripheral blood lymphocyte cultures (11). Add $0.05-0.06$ μg/ml colcemid (Gibco) to the culture and incubate for $10-15$ min. (If the cells have not been synchronised, incubate with colcemid for 20 min to 2 h.)

(ii)   Transfer the cells to a centrifuge tube and centrifuge at 200 *g* in a clinical centrifuge for 8 min. Remove most of the supernatant and resuspend the cells in the residual 0.5 ml of medium.

(iii)   Add freshly-made 75 mM KCl at 37°C. Gently resuspend the cells. Incubate for 10 min at 37°C.

(iv)   Centrifuge for 5 min at 200 *g* in a clinical centrifuge. Remove almost all of the supernatant. Then flick the tube a couple of times with a finger to loosen the pellet.

(v)   Add 1 ml of freshly-made methanol:acetic acid (3:1 v/v) drop by drop while shaking the tube. Mix thoroughly and then add more of this fixative to 5 ml final volume.

(vi)   Cap the tube and leave at room temperature for 20 min. Recover the cells by centrifugation.

(vii)   Resuspend the cells in 2 ml of fresh fixative. Centrifuge the cells down. Repeat this rinse 1 − 4 times more.

(viii)   Resuspend the cells in fresh fixative and drop this onto slides that have been pre-cleaned with 70% ethanol. The best conditions for spreading will depend on the humidity. If the atmosphere is dry, drop the cells onto ice-cold, wet slides. If the atmosphere is humid, drop the cells onto dry slides at room temperature.

(ix)   Check the preparations using a phase-contrast microscope. The best preparations have cells with no refractile qualities. The slides should be used for hybridisation within 2 weeks of preparation.

4.1.3 *Small Pieces of Tissue*

(i)   Tease freshly dissected tissue into very small pieces, not more than 5 mm in their longest dimension. Fix for 5 − 10 min in freshly mixed ethanol:acetic acid (3:1 v/v).

(ii)   Place a small drop of 45% acetic acid onto a siliconised coverslip (18 mm²) prepared as described in Section 2. Tease a small piece of tissue (< 1 mm²) from the fixed tissue and transfer it to the drop on the slide. Addition of the fixative to the drop of acid will cause violent mixing. This can be prevented by holding the tissue briefly in air to allow evaporation of the fixative before placing it in the acid. However, the morphology may be ruined if the tissue becomes too dry at this stage.

(iii)   Thoroughly mince the tissue in the drop of acetic acid and remove any remaining large pieces.

(iv)   Carefully lower a subbed slide onto the coverslip and squash the cells. A simple way to squash the cells is to place the slide on filter paper with the coverslip side down and apply firm thumb pressure on the back of the slide over the coverslip. Some tough tissues will give better preparations if the slide is left on a warming plate at 45°C for 3 − 5 min to allow the tissue to soften before being squashed. The flattest squashes have the best preservation of morphology during the hybridisation steps so it is wise to use small pieces of tissue and to remove any large pieces that might interfere.

(v)   Immediately after squashing, place the slide coverslip up on a flat piece of dry ice.

(vi) When the preparation is completely frozen (5 – 10 min), flip off the coverslip by inserting a razor blade under one corner.

(vii) Immediately plunge the slide into 95% ethanol at room temperature. Leave for 10 min.

(viii) Put the slide through two additional washes of 95% ethanol (10 min each). Allow the slide to air-dry. Slides can be stored dry for long periods. Good preparations of tissue are very flat; when examined dry with a phase-contrast microscope they show good morphology and have no refractile areas.

4.2 Hybridisation - Standard Protocol

If the cytological preparation is to be examined only by *in situ* hybridisation, the protocol for hybridisation is as described in *Table 4*. For mammalian metaphase chromosomes that are to be identified by G-banding, a modified procedure must be used (see Section 4.3). The rationale for several of the steps involved may not always be immediately obvious and therefore additional notes are given below.

(i) The author has found that a heat treatment followed by dehydration using ethanol (*Table 4*, steps 1 and 2) significantly reduces the loss of morphology that results when part of a chromosome does not tightly adhere to the slide (12).

(ii) In early studies on rRNA genes in polytene chromosomes and in oocytes, it was found that RNase treatment significantly improved the specific hybridisation of labelled rRNA used as a probe (13). Presumably the RNase treatment removes endogenous rRNA which might act as a competitor for the labelled probe.

Table 4. Hybridisation to DNA in Cytological Preparations.

Pre-treatment of Slides before Hybridisation

1. Incubate the slides[a] in 2 x SSC[b] at 70°C for 30 min.
2. Transfer the slides to 70% ethanol at room temperature for 10 min. Wash again in 70% ethanol for 10 min and then in 95% ethanol for 5 min. Air dry.
3. Arrange the slides in a moist chamber containing 2 x SSC. Place 100 μl of RNase[c] (100 μg/ml in 2 x SSC) over the preparation on each slide and cover each with a coverslip (22 mm²).
4. Incubate the slides at room temperature for 2 h. Then remove the coverslips gently by dipping the slides into a beaker of 2 x SSC to float off the coverslips.
5. Wash the slides in 2 x SSC (3 x 5 min), 70% ethanol (2 x 10 min) and 95% ethanol (5 min.) Air dry.
6. Suspend the slides in 0.1 M triethanolamine-HCl, pH 8.0. Stir the solution vigorously and add acetic anhydride to 5 ml per litre.
7. When the acetic anhydride is thoroughly dispersed, stop the stirring and leave for 10 min.
8. Wash the slides in 2 x SSC (5 min), 70% ethanol (2 x 10 min) and 95% ethanol (5 min). Air dry.
9. Place the slides in 70 mM NaOH for 3 min. Then wash the slides in three changes of 70% ethanol (10 min each) and two changes of 95% ethanol (5 min each).
10. Air dry the slides.

Hybridisation

11. Hybridisation is performed *either* in aqueous salt solution such as TNS buffer (0.15 M NaCl, 10 mM Tris-HCl, pH 6.8) at high temperature *or* in the presence of formamide at lower temperature.

Hybridisation in aqueous salt solution at high temperature.

This is carried out typically in 0.3 M NaCl, 20 mM Tris-HCl, pH 6.8, at 67°C. The probe is generally used at 1 – 10 ng per slide. If a DNA probe is being used, sheared, denatured *E. coli* DNA is also added at 4.0 μg per slide to reduce non-specific binding of the probe. If an RNA probe is being used, *E. coli* rRNA is added instead.

(i) If the probe is double-stranded DNA or double-stranded complementary RNA (cRNA), it should be denatured immediately before use as follows. Mix the probe with the carrier *E. coli* DNA or *E. coli* rRNA in water to 90% of the final volume required to give the correct concentrations. Incubate the mixture at 85°C for 3 − 15 min and then chill quickly. Bring to 0.3 M NaCl, 20 mM Tris-HCl, pH 6.8, by adding 1/10th volume of 10 x concentrated stock buffer. Single-stranded probes such as M13 clones or SP6 transcripts do not need to be denatured and so can be dissolved directly in 0.3 M NaCl, 20 mM Tris-HCl, pH 6.8, together with the carrier DNA or rRNA.

(ii) Place 15 − 20 μl of the hybridisation mixture over each cytological preparation and cover it with a coverslip (18 mm²).

(iii) Place the slides over a reservoir of hybridisation buffer in a tightly-sealed moist chamber[e] in an oven at 67°C. Note that the amount of liquid in the chamber must be sufficient to prevent evaporation from under the coverslip, otherwise the autoradiographs will show a high background. Typical hybridisation times are 12 − 14 h at 67°C.

Hybridisation in formamide solution.

This is carried out typically in 40% formamide[d] in 4 x SSC[b] at 40°C. The amounts of probe and *E. coli* DNA used per slide are as for hybridisation in aqueous salt solution (see above).

(i) If the probe is double-stranded DNA or double-stranded cRNA, it should be denatured immediately before use. To do this, dissolve the probe plus carrier *E. coli* DNA or rRNA in water to 40% of the final volume desired. Heat the mixture at 85°C for 3 − 15 min and then chill quickly. Add 20% of the final volume of 20 x SSC and 40% of the final volume of formamide. Single-stranded probes such as M13 clones or SP6 transcripts do not need to be denatured and so can be dissolved directly in 40% formamide, 4 x SSC plus carrier *E. coli* DNA or rRNA.

(ii) Place 5 μl of the hybridisation mix over each preparation and cover with a coverslip (18 mm²).

(iii) Seal the coverslip with a thick coat of rubber cement. The seal saves the expense of filling a moist chamber with formamide buffer to prevent evaporation from under the coverslip.

(iv) Incubate the sealed slides in a closed container on moist paper towels[e] to prevent drying of the rubber cement. Typical incubations are for 12 − 14 h at 40°C.

Treatment of Slides after Hybridisation

12. Following hybridisation, remove the slides from the moist chambers. Immediately wash off the coverslip and hybridisation mixture by dipping the slide into a beaker of 2 x SSC. (If the coverslip is sealed, peel off the rubber cement first.) Place the slides in a rack in a staining dish containing 2 x SSC and leave for 15 min.

13. Remove non-specifically bound nucleic acid probe as follows:

RNA probes

(i) Place the slides in RNase[c] (20 μg/ml in 2 x SSC) at 37°C for 1 h.

(ii) Rinse the slides twice in 2 x SSC for 10 min each time.

(iii) Dehydrate the preparation in 70% ethanol (2 x 10 min) and 95% ethanol (5 min). Air dry.

DNA probes

(i) Remove non-specifically bound DNA by three additional 10 min washes in 2 x SSC at a temperature 5°C less than the temperature used for the hybridisation reaction.

(ii) Dehydrate the preparations in 70% ethanol (2 x 10 min) and 95% ethanol (5 min). Air dry.

14. The preparations are now ready for autoradiography (see Section 6).

---

[a] Slides bearing the tissue to be examined by *in situ* hybridisation are prepared by the methods described in Section 4.1.

[b] The composition of SSC is 0.15 M NaCl, 0.015 M trisodium citrate, pH 7.0 (with NaOH).

[c] Pancreatic RNase is dissolved at 1 mg/ml in 20 mM sodium acetate, pH 5.0, and placed in a boiling water bath for 5 min. Aliquots are stored at −20°C and diluted as required immediately before use. Pre-treatment of cytological preparations with RNase to remove total RNA is carried out with a high enzyme concentration (100 μg/ml). Post-treatment of preparations after hybridisation to remove only non-hybridised RNA is carried out with a lower concentration of enzyme (20 μg/ml).

[d] Deionised formamide should be used. This is prepared as follows. Mix 100 ml formamide with 5 g mixed-bed ion-exchange resin (e.g., BioRad AG501-X8, 20 − 50 mesh). Stir at room temperature for 30 min or until the colour of the resin changes. Filter through Whatman filter paper to remove the resin. Repeat twice more. Store aliquots of the deionised formamide at −70°C.

[e] See Section 2.4.

Although the effect of RNase treatment on the hybridisation of other DNA sequences has not been tested, it seems sensible to include RNase treatment if there is any reason to believe that there might be a significant concentration of RNA transcripts near the gene. In practice, therefore, it is wise to treat all cytological preparations routinely with RNase (*Table 4*, steps 3 − 5) prior to hybridisation. Careful washing after the RNase treatment will ensure that no enzyme remains to degrade the hybridisation probe.

(iii) Steps 6 − 8 are intended to acetylate the cytological preparation. This has proved to be a most effective way to reduce non-specific binding of negatively-charged nucleic acid probes (14).

(iv) The DNA in the cytological preparation must be denatured prior to exposure to the probe or hybridisation will not occur. All of the agents that denature purified DNA, such as alkali, heat, and organic solvents, also denature DNA in cytological preparations as measured by *in situ* hybridisation. Acid treatments also denature DNA on slides but give a reduced amount of *in situ* hybridisation. This is because the acid treatments that denature DNA in cytological preparations also produce some depurination even though chromosome morphology is excellently preserved during acid denaturation (M.L.Pardue and H.C.MacGregor; unpublished data). In the author's experience, alkali denaturation (*Table 4*, step 9) consistently gives the best results.

(v) Most studies of the parameters affecting *in situ* hybridisation have indicated that similar considerations apply as for hybridisation to DNA fixed on nitrocellulose filters. Therefore conditions of ionic strength, temperature, probe concentration and hybridisation time are chosen for each experiment in the same way as one would choose them for filter hybridisation experiments (see Chapter 3). The conditions chosen determine the stringency of the hybridisation reaction and set limits on the amount of sequence mismatching that can occur. *In situ* hybridisation can be carried out either in salt solution at high temperature or in formamide solution at lower temperature. Protocols for moderate stringency hybridisation under these two conditions are given in *Table 4, step 11*.

Dextran sulphate has been shown to accelerate the rate of nucleic acid hybridisation both in solution (15) and to immobilised nucleic acids (16). The significant acceleration seen for the hybridisation of randomly-cleaved double-stranded probes appears to be due to the formation of networks of probe on the target. The hybridisation of single-stranded probes to immobilised DNA is also accelerated but less so than for double-stranded probes. In this case, dextran sulphate is thought to act by concentrating the probe by exclusion from the volume of the solution occupied by the polymer. If desired, either of the compositions of the hybridisation mixtures given in *Table 4*, step 11 can be adjusted to include 10% dextran sulphate.

(vi) A word about coverslips. Some brands of coverslips leach alkali into the hybridisation solution during incubation at typical hybridisation temperatures. This results in a significant increase in the pH of the solution under the coverslip and prevents hybrid formation. Therefore it is wise to test a new brand of coverslip before using it. This can be done by performing a 'mock' hybridisation using hybridisation buffer plus phenol red under the coverslip. If the indicator shows

Table 5. Hybridisation to DNA in Mammalian Metaphase Chromosomes.

Removal of Endogenous RNA

1. Treat the preparations with RNase[a] (100 μg/ml in 2 x SSC[b]) at 37°C for 1 h.
2. Rinse in 2 x SSC (3 x 5 min).
3. Dehydrate in 70% ethanol for 2 x 10 min then 95% ethanol for 5 min. Air dry.

Denaturation of Chromosomal DNA

4. Immerse the slides in 70% (v/v) formamide in 2 x SSC at 70°C for 2 min.
5. Dehydrate in ethanol as in step 3.

Hybridisation

6. Apply 20 μl of labelled probe at 0.05 − 0.2 μg/ml in 50% (v/v) formamide in 0.3 M NaCl, 30 mM trisodium citrate, 10% dextran sulphate, 40 mM sodium phosphate buffer, pH 6.0, containing a 1000-fold excess of sheared, denatured non-competing DNA[c].
7. Cover the hybridisation solution with a coverslip (18 mm^2) and seal thickly with rubber cement.
8. Incubate at 37°C for 11 h.

Removal of Non-specifically Bound Probe

9. Rinse the slides in 50% formamide in 2 x SSC at 39°C.
10. Rinse the slides in 2 x SSC at 39°C.
11. Dehydrate the preparations in ethanol as in step 3. Air dry.
12. The preparations are now ready for autoradiography (Section 6).

---

[a]See *Table 4,* footnote c
[b]See *Table 4,* footnote b
[c]See *Table 3,* footnote d

that the coverslips are leaching alkali during the incubation (indicator turns purple), the coverslips should be boiled in 1 M HCl and rinsed well in 10 mM phosphate buffer, pH 7.0, before routine use.

4.3 Hybridisation to DNA in Human Metaphase Chromosomes

For mammalian metaphase chromosomes that are to be identified by G-banding after hybridisation, the standard hybridisation procedure (*Table 4*) is modified (17,18) as described in *Table 5*.

5. HYBRIDISATION TO RNA IN CYTOLOGICAL PREPARATIONS

5.1 Cytological Procedures

Most hybridisations to cellular DNA require only the preservation of chromosome or, at most, nuclear morphology. In contrast, most studies of the localisation of RNA in cytological preparations require preservation of the morphology of the entire cell and, in some cases, of tissues or organs. It would be expected, therefore, that techniques to make cytological preparations for RNA studies would be even more cell type-specific than the techniques used for DNA studies. Examples of procedures that give excellent results with different types of biological materials are given in references 4, 19 − 22. Many other conventional techniques for fixation should be compatible with hybridisation to cellular RNA so it is worthwhile to experiment with several techniques if one is studying a different type of biological material. The best cytological techniques for hybridisation to cellular RNA should:

(i) preserve the morphology of the cells;
(ii) not extract or modify the RNA;

(iii)   not change the localisation of the RNA;

(iv)   leave the RNA accessible to the hybridisation probe.

Formaldehyde fixatives, which are not used for DNA hybridisation because they appear to interfere with DNA denaturation, do give good results in some RNA hybridisation procedures. However, formaldehyde-fixed tissues embedded in methacrylate give reduced *in situ* hybridisation. This appears to be a result of an interaction between the formaldehyde and the methacrylate since both reagents have given good results in other combinations (20). The possibility of such interactions should be kept in mind when testing other techniques. Three procedures are described below (Sections 5.1.1—5.1.3) as examples of techniques that can be used for the preparation of cytological specimens for studies of cellular RNA. An example of a preparation made by the freeze-substitution and methacrylate embedding procedure is shown in *Figure 4*.

In most experiments using hybridisation to RNA in cytological preparations, hybridisation to the DNA of the cell does not complicate the results. There are several reasons for this. The DNA is localised in the nucleus while most of the RNA studied is cytoplasmic. Protocols for hybridisation to RNA in cytological preparations do not denature DNA. When the RNA being studied is nuclear, it is usually much more abundant than the corresponding DNA sequence since one gene can produce many transcripts. However, there are some cases, for example puffs on polytene chromosomes, where the ratio of RNA to DNA is not high and where a significant amount of single-stranded DNA is found after gentle cytological procedures. If there is a possibility that the probe might bind significantly to DNA rather than RNA, some slides should be treated with RNase, as in *Table 4*, steps 3 and 4, prior to hybridisation. The detection of autoradiographic grains in such RNase-treated slides indicates the level and location of any hybridisation to DNA in the preparation.

5.1.1 *Freeze-substitution and Methacrylate Embedding of Drosophila Ovaries* (20)

(i)   Fix very small pieces of tissue by plunging them into absolute ethanol held at the temperature of dry ice. After 24—36 h at this low temperature, allow the dry ice to sublime so that the ethanol comes slowly to room temperature.

(ii)   Transfer the tissue to a 1:1 mixture of ethanol and methacrylate monomer for 30 min. (Methacrylate monomer is prepared by mixing 9 parts butyl methacrylate and 1 part methyl methacrylate. This mixture keeps well at room temperature in a brown bottle.)

(iii)   Transfer the tissue to methacrylate monomer for 30 min.

(iv)   Fill an electron microscope embedding capsule with methacrylate monomer containing 1% benezoyl peroxide (freshly prepared). Place the tissue in the medium and seal the capsule tightly since polymerisation of the methacrylate is inhibited by air. Leave at 60°C for 18—24 h.

(v)   Cut sections (1 μm thick) of the embedded tissue.

(vi)   With a fine brush, transfer the sections to a small drop of distilled water on a subbed slide (see Section 2).

(vii)   Hold a swab or piece of cardboard saturated with chloroform over the sections briefly to allow the vapour to spread the sections.

(viii)   Pull the excess water off the slide with a piece of filter paper. Allow the sections to dry by placing the slide on a warming plate at 60°C.

(ix)   Swirl the slide in a glass beaker containing xylene to dissolve the embedding resin.

(x)   Rehydrate the preparation by transferring the slide to the following solutions in order (5 min each); xylene:absolute ethanol (1:1 v/v), absolute ethanol, 95% ethanol, 70% ethanol, 30% ethanol, water. Leave the sections in water until the hybridisation probe is ready to be applied (*Table 6*).

5.1.2 Glutaraldehyde Fixation and Paraffin-embedding of Sea Urchin Eggs and Embryos (4,19)

(i)   Fix the eggs or embryos in 1% (v/v) glutaraldehyde, 0.43 M NaCl, 50 mM sodium phosphate, pH 7.5, at 0°C for 60 min.

(ii)   Wash in buffer at 0°C (2 x 30 min).

(iii) Dehydrate the tissue by incubation (30 min each time) in increasing concentrations of ethanol [50%, 70%, 85%, 95%, 99%, 99%, absolute ethanol:xylene (1:1 v/v)].

(iv)   Incubate the tissue for 30 min in xylene (twice), 45 min in xylene:paraffin (m.p. 55 − 57°C; 1:1 v/v) at 55°C, and then for 5 min (three times) in melted paraffin (m.p. 55 − 57°C) at 58°C. Place small drops of the paraffin containing the fixed tissue in embedding capsules and allow them to solidify. Heat the paraffin until it just begins to glisten and then cover it with liquid paraffin. These paraffin blocks can be stored at 4°C for several days at least.

(v)   Cut sections (5 μm thick) and place these on subbed slides. Dry for at least 48 h on a warming plate set at 40°C.

(vi)   Remove the paraffin by washing in xylene (2 x 10 min), and rehydrate the tissue by washing (5 min each) in decreasing concentrations of ethanol and finally in water.

5.1.3 Preparation of Frozen Sections of Fixed Drosophila Embryos (21,22)

(i)   Harvest embryos of the desired age onto nylon mesh.

(ii)   Dechorionate the embryos by immersion in 3% sodium hypochlorite for 2 min. Wash with phosphate-buffered saline (PBS) containing 0.13 M NaCl, 7 mM Na_2HPO_4, 3 mM NaH_2PO_4. Blot dry.

(iii) Collect the embryos using a paint brush, and transfer them to the fixative. (This is prepared by mixing equal volumes of n-heptane and *freshly made* 4% para-formaldehyde in PBS. Shake for 1 min at room temperature, allow the phases to separate and then recover the organic phase. This is the phase used.) Incubate the embryos in the fixative for 10 min at room temperature with shaking. This permeabilises the vitelline membrane.

(iv)   Drain onto a filter and immediately transfer the filter with the eggs to methanol/n-heptane [9 ml methanol, 1 ml 0.5 M EGTA, pH 8.0, 10 ml n-heptane] pre-cooled in dry ice/ethanol. Shake vigorously for 10 min.

(v)   Heat the tube rapidly to room temperature by swirling under hot running water.

(vi)   Allow the devitellinised embryos to settle to the bottom of the tube and collect them using a Pasteur pipette.

(vii) Transfer the embryos to 5 ml of 90% methanol, 50 mM EGTA, pH 8.0, in a 10 ml conical tube.

(viii) Wash, post-fix and rehydrate the embryos by sedimenting gently in a 10 ml conical bottom tube in the following solutions (5 ml each, room temperature):

90% methanol, 50 mM EGTA

repeat

90% methanol, 50 mM EGTA:4% paraformaldehyde in PBS (7:3 v/v)

repeat

90% methanol, 50 mM EGTA:4% paraformaldehyde in PBS (5:5 v/v)

repeat

90% methanol, 50 mM EGTA:4% paraformaldehyde in PBS (2.5:7.5 v/v)

repeat

4% paraformaldehyde in PBS

PBS

repeat

(ix) Transfer the embryo pellet to OTC embedding medium (Miles Laboratories) and mount on a cryostat specimen holder. Freeze in liquid nitrogen.

(x) Prepare 8 μm sections and place on polylysine-coated slides (see Section 2.3) at room temperature.

(xi) Allow the sections to melt and dry flat on the slides.

(xii) Heat the slides in a slide warmer at 50°C for 1−2 min. Allow to cool. The slides can now be left for 1−2 h at room temperature if this is more convenient.

(xiii) Post-fix the tissue by immersing the slides in *fresh* 4% paraformaldehyde in PBS for 20 min at room temperature.

(xiv) Wash the slides in PBS at room temperature for 5 min.

(xv) Dehydrate by immersion (2 min each) in the following series of ethanol concentrations; 30%, 60%, 80%, 95%, 100%.

(xvi) Allow the slides to air-dry. The preparations are now ready for pre-hybridisation treatments (see Section 5.2).

5.2 Hybridisation

There is no standard procedure suitable for all tissues and all applications. A general guide to the technique is given in *Table 6* and additional notes are given below.

(i) Cytological preparations are sometimes treated prior to hybridisation to improve access of the cellular RNA to the probe (*Table 6*, step 1). The most commonly used method is treatment with proteolytic enzymes. Paraffin-embedded sea urchin embryos have been treated with 1 μg/ml Proteinase K in 50 mM EDTA, 0.1 M Tris-HCl, pH 8.0, for 30 min at 37°C, followed by washing in distilled water (4,19). On the other hand, no pre-treatments were used for *Drosophila* ovaries embedded in methacrylate (20).

(ii) Acetylation of the preparation is an effective way to reduce non-specific binding of the nucleic acid probe (*Table 6*, step 2). This has proved useful in all types of *in situ* hybridisation analyses and is therefore routinely used by the author.

(iii) The probe can be either RNA or DNA. Since the target is single-stranded RNA, the most effective method is to use a single-stranded probe complementary to the target RNA. This is most easily prepared by constructing an appropriate SP6 recombinant which will then produce anti-sense RNA (see Section 3.3.2).

(iv) Hybridisation to cellular RNA is usually done in formamide to prevent exposing

Table 6. Hybridisation to RNA in Cytological Preparations.

Pretreatment of Slides

1.   The cytological preparation may first be digested with protease to improve access of the probe to cellular RNA [see Section 5.2 (i)].
2.   Acetylate the preparation (as in *Table 4*, steps 6−9) to reduce non-specific binding of the probe. Alternatively, wash the preparation with physiological buffer (the composition will depend on the tissue under study), transfer to distilled water and use immediately for hybridisation. In this case, take care that no water is left on the slide to cause dilution of the probe.

Hybridisation

3.   Hybridisation can be carried out in 40% formamide[a], 4 x SSC[b], 0.1M sodium phosphate buffer, pH 8.0, at 50°C. Somewhat more stringent conditions are obtained in 50% formamide, 5 mM EDTA, 0.3 M NaCl, 20 mM Tris-HCl, pH 8.0 at 50°C. The probe is generally used at 1−10 ng per slide. For DNA probes, sheared denatured *E. coli* DNA[c] is added at 4.0 μg per slide to reduce non-specific binding of the probe. For RNA probes, *E. coli* rRNA is used instead.

    (i)   Prepare the hybridisation mixture containing the labelled probe and *E. coli* DNA or rRNA (see above) at the required concentrations. Details of probe preparation are given in *Table 4*, (step 11). Denhardt's solution [0.02% each of bovine serum albumin (BRL), Ficoll (Sigma), and polyvinylpyrollidone (Sigma)] can also be added to reduce the background. As with DNA hybrids [Section 4.2 (v)], dextran sulphate can be added to accelerate hybridisation.

    (ii) Place 5 μl of the hybridisation mixture over each preparation and cover with a coverslip (18 mm²). Seal the coverslip with a thick coat of rubber cement (see *Table 4*, step 11). In some cases, after methacrylate embedding (Section 5.1.1), dry preparations do not wet well with the formamide solution. Therefore when using the methacrylate procedure, leave the slides in water until just before the probe is applied. Then wipe each slide and flip it hard with a finger to remove remaining drops of water. Apply the probe before the preparation dries completely. Cover each preparation with a coverslip and seal with rubber cement as before.

    (iii) Incubate the slides in a closed chamber on moist paper towels [d] at the chosen temperature (typically 50°C). Typical incubations are for 12−14 h.

Treatment of Slides after Hybridisation

4.   Following hybridisation, remove the slides from the moist chambers, peel off the rubber cement and rinse off the coverslip and hybridisation mixture by dipping each slide into a beaker of 2 x SSC. Place the slides in a staining rack in a dish containing 2 x SSC and leave for 15 min.
5.   Remove non-specifically bound probe as follows:

RNA probes

    (i)   Place the slides in RNase[e] (20 μg/ml) in 0.5 M NaCl, 10 mM Tris-HCl, pH 8.0, at 37°C for 30 min.

    (ii) Wash the slides in this buffer at 37°C for 30 min, in 0.1 x SSC at 50°C for 10 min, and finally in 0.1 x SSC at room temperature for 10 min.

    (iii) Dehydrate the preparations in 70% ethanol (2 x 10 min) and 95% ethanol (5 min). Air dry.

DNA probes

    (i)   Remove the non-hybridised probe by repeated washings in the hybridisation buffer at a temperature a few degrees below the temperature of hybridisation.

    (ii) Wash the slides at room temperature in 0.1 x SSC.

    (iii) Dehydrate the preparations in 70% ethanol (2 x 10 min) and 95% ethanol (5 min). Air dry.

6.   The preparations are now ready for autoradiography (Section 6).

[a]See *Table 4*, footnote d
[b]See *Table 4*, footnote b
[c]See *Table 3*, footnote d
[d]See Section 2.4.
[e]See *Table 4*, footnote c

the tissue to the higher temperatures necessary for RNA-RNA hybrids in salt solutions. [Both naturally occurring RNA duplexes and synthetic polymers have melting points (T_ms) as much as 16°C above those of the corresponding DNA (23,24).] In calculating an appropriate temperature for hybridising an RNA probe to cellular RNA, it should be noted that formamide reduces the melting point of RNA-RNA duplexes by 0.35°C for every 1% formamide rather than the 0.65°C per 1% formamide obtained with DNA-DNA duplexes (4). Two sets of hybridisation conditions, which differ in stringency, are given in *Table 6*, step 3.

(v)   Because hybridised RNA is relatively resistant to digestion by pancreatic RNase, it is possible to use mild RNase treatment (*Table 6*, step 5) to remove non-specifically bound RNA probes.

6. AUTORADIOGRAPHY

6.1 Introduction

As stated earlier, the most useful isotope for cytological autoradiography is tritium which emits weak β-radiation. The half-life of tritium is 12.5 years and the mean β particle energy is 0.018 MeV. In biological material the approximate range of the tritium β particle is generally less than 1 μm although $1-2\%$ of the β particles may travel up to 3 μm. The β particles are detected by thin films of autoradiographic emulsion that adhere very closely to the surface of the cytological preparation.

6.2 Equipment and Reagents

6.2.1 The Darkroom

Autoradiography requires a completely light-tight darkroom. Since slides must be left to dry for at least 2 h in the dark after application of the autoradiographic emulsion, it is useful for the darkroom to have a double light-tight door to allow the experimenter to leave without admitting light into the room. If there is no double door the slides can be dried in a well-ventilated light-tight box.

A glass or plastic slide dipping chamber is needed which should have a cavity of approximately 3.0 cm x 0.6 cm and 6.0 cm deep so that a slide held by one end can be dipped to the bottom. This will hold approximately 10 ml of emulsion. A simple chamber can be made by cutting the top off a polyethylene two-place slide mailer sold by medical supply houses. A wire test-tube rack can be used for drying slides. Slides can stand on end diagonally in each of the square test tube compartments.

The darkroom should also have a constant temperature water bath set at $43-45$°C, a rack for drying slides and a photographic safelight. For the autoradiographic emulsion used in these experiments the safelight should have a Kodak Wratten Series II or 0A filter and be fitted with a 25 W bulb. The safelight should be kept as far from the working area as possible. If the working area is carefully arranged, it is not difficult to carry out the necessary operations with little or no light.

6.2.2 The Autoradiographic Emulsion

The emulsion used for *in situ* hybridisation is Kodak NTB-2 nuclear track emulsion (Eastman Kodak Co; DC Special Products Division). It is solid at room temperature and becomes liquid when warmed to 45°C. For autoradiography of tritium, use the emulsion diluted with an equal part of distilled water. This dilute emulsion produces a thin, even coating on the slide. A thicker coating of emulsion on the slide will not increase the information obtained since the β-particles emitted by tritium are not energetic enough to reach the upper part of the film. On the other hand, a thicker film has more emulsion where background grains can be produced by extraneous causes. In addition to improving the signal-to-noise ratio, thinner films also improve the optical qualities of the preparation.

Kodak NTB-2 should be purchased in 4 oz bottles (112 ml) and stored in the refrigerator. It is important that all emulsion and exposing slides be kept in a refrigerator that is never used for other radioisotopes. Even a very dilute ^{32}P-solution in the same refrigerator will rapidly produce unacceptable background. Each new batch of emulsion should be diluted and stored in aliquots of approximately 10 ml, enough for single use, as follows:

(i)   The NTB-2 package is opened in the darkroom under the safelight and the 4 oz bottle is warmed at 45°C for 30 min. At the same time a 500 ml Erlenmeyer flask containing 112 ml distilled water is also warmed to 45°C.

(ii)   After the NTB-2 has melted it is poured very slowly down the side of the flask which is held at an angle. Any bubbles produced in this mixing step will persist and invariably end up on top of a preparation. The emulsion and water mix easily. One or two slow rotations of the flask, still held at an angle, are sufficient for mixing.

(iii)   The diluted emulsion is then distributed into nylon scintillation vials in aliquots of about 10 ml (enough to dip ~ 30 slides if the tissue preparations have been placed close to the bottom of the slide). The vials are wrapped in aluminum foil and stored in a light-tight box in the refrigerator. The triple boxes in which Kodak electron microscope plates are supplied are ideal for storing vials of emulsion as well as for storing the plastic boxes of slides undergoing autoradiography. If the caps on the scintillation vials have cork inserts, these should be removed since they may give off organic compounds which can produce background grains in the emulsion during storage.

Diluted emulsion will keep indefinitely in the refrigerator if it is not exposed to radioactivity or fumes of organic chemicals. Emulsion remaining after a batch of slides has been dipped can be stored and reused; however the background should be checked to be sure that it is still acceptable.

The author tests each new batch of NTB-2 at the time it is diluted and re-packaged. Two clean subbed slides (Section 2.1) are dipped and dried in the dark for 20 min. They are developed and a drop of water is put on each slide and covered with a coverslip. The slide is examined microscopically using a 100 x Neofluar lens and 10 x Kpl eyepieces. If the average number of grains per microscope field is less than 100 the emulsion is acceptable. If it is less than 50 the emulsion is very good. If the emulsion arrives from Kodak with much more than 100 grains per field, the author does not accept it.

6.3 Autoradiographic Procedure

(i)   Place the dipping chamber and a vial of emulsion in a 45°C water bath for 10 – 15 min. Then pour the emulsion slowly down the side of the chamber so that no bubbles form.

(ii)   One at a time, dip the slides in slowly and withdraw. Drain the slides briefly and place them on end in a slide rack.

(iii)   Allow the slides to dry for at least 2 h. A non-sparking fan can be used to reduce this time to 1 h.

(iv)   Store the dried slides in light-tight plastic boxes together with a small container of silica gel to maintain dryness. Seal the boxes with black electrical tape and store these in light-proof containers in the refrigerator for exposure to occur. Moisture will cause fading of the latent image and with very weak signals the fading of the image can be nearly equal to its production. However, if the slides are kept very dry even very weak signals can be detected after a long exposure. The autoradiographic exposure time may range from a few hours to many days. Therefore it is necessary to have several replicate slides so that test slides can be developed at intervals to determine the proper exposure time.

(v)   Although the exact temperature is not critical, it is best to develop the autoradiographs at $15-20°C$. The lower the temperature, the smaller the grain size. However, it is more important that the slides and all of the developer and fixer solutions are at the same temperature since temperature changes may produce wrinkles in the emulsion. The schedule for developing autoradiograms is:

>   Kodak D-19 developer: 2.5 min
>   Stopbath (water):   30 sec
>   Kodak rapid fixer (undiluted):   5 min
>   Distilled water: 5 min wash repeated a total of 5 times. This step can be carried out in daylight.

While the slides are in the developer, agitate them gently by rocking the dish back and forth for cycles of 6 sec movement and 6 sec rest. The movement ensures that the developer is mixed and does not become depleted near areas of high grain density on the slides. Slides should also be agitated gently in the other solutions from time to time. However, the amount of agitation is not as critical as during the incubation in the developer.

(vi)   After the final rinse in distilled water, air dry the slides or stain them immediately.

7. STAINING PREPARATIONS AFTER AUTORADIOGRAPHY

7.1 **Giemsa Staining**

There are a number of cytological stains that will stain preparations through an autoradiographic emulsion. One of the most useful is Giemsa stain which works well for both chromosome preparations after DNA hybridisation and for tissue sections after RNA hybridisation.

(i)   Immediately before use, prepare the stain by diluting a stock solution of Giemsa blood stain (Harelco) 1:20 with 10 mM phosphate buffer, pH 6.8.

(ii)   Dip the slides in the stain. The time required for staining (between 2 and 45 min) depends both on the cytological preparation and on the batch of Giemsa. A metallic film forms over the surface of the Giemsa solution soon after it has been mixed with buffer [step (i)]. Since this film will adhere to the slide if the slide is pulled out of the solution, distilled water is added to overflow the dish

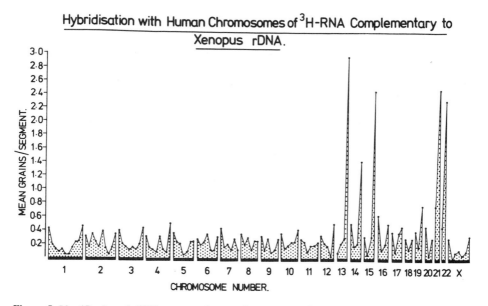

Figure 5. Identification of rRNA genes on human chromosomes. Chromosomes were identified by their Q-banding pattern (e.g., as in *Figure 2*). Since chromosome contraction may take place to different extents in individual cells and must therefore be corrected for, each chromosome was divided into a number of equal-sized segments. The standard segment was half the length of the smallest autosome. Grains were then counted from the autoradiograph of the chromosomes. The numbers of grains were totalled for each segment and divided by the number of segments counted, to give a mean grain count for each segment. (See reference 1 for details). The rRNA hybridises to the five nucleolar organiser regions in the human chromosome set.

and float away the film before the slides are removed. The progress of staining can be checked by removing the slide from the stain, rinsing it with distilled water and viewing it under the light microscope while it is still wet. If the slide is understained it can be put back in the stain. Overstained slides can be destained in ethanol but the morphology is never as good after destaining. The aim is to stain autoradiographs only lightly so that autoradiographic grains are clearly visible in black and white photographs.

(iii) After Giemsa staining, rinse the slides briefly in distilled water and allow them to air-dry for at least 1 h.

(iv) Mount each stained preparation under a coverslip with a small drop of Permount. Alcohol-based mounting media, such as Euparol, should not be used since they will eventually destain the preparation. Alternatively, leave the autoradiographs uncovered and view them under immersion oil. Use petroleum ether to remove oil from unmounted slides.

7.2 G-banding for Human Metaphase Chromosomes

Human chromosomes can be identified unambiguously only if examined using one of the banding procedures. Preparations can be Q-banded (i.e. using quinacrine), photographed and then hybridised. The autoradiograph is then compared with the Q-band photograph for quantitative analysis as in *Figures 2* and *5*. Alternatively, recent modifications permit G-banding of chromosomes through the autoradiographic emulsion (17,18) as follows.

(i)   After autoradiography, stain the chromosomes in Wright's stain diluted 1:3 with 60 mM phosphate buffer, pH 6.8, for 8 − 10 min (25).

(ii) The staining contrast can be enhanced by destaining and restaining once or twice according to the following cycle: 95% ethanol (2 min), chloroform (15 sec), 95% ethanol plus 1% HCl (30 sec), 100% methanol (2 min), Wright's stain (6 − 8 min).

7.3 Coomassie Blue Staining

Preparations that stain very faintly with Giemsa, such as lampbrush chromosomes and very flattened cells, can be stained with Coomassie Blue.

(i)   Pass the air-dried slides quickly through 50% methanol into 0.1% Coomassie Blue R250, 50% methanol, 10% acetic acid.

(ii) After 10 min in this stain, dip the slides into 50% methanol, then wash for 10 min in water and air dry.

8. ACKNOWLEDGEMENTS

I am grateful to J.G.Gall, M.E.Harper, K.Valgeirsdottir, W.Bendena, A.Roter and K.Traverse for helpful discussions. My work has been supported by grants from the National Institutes of Health and from the American Cancer Society.

9. REFERENCES

1. Evans,H.J., Buckland,R.A. and Pardue,M.L. (1974) *Chromosoma (Berl.)*, **48**, 405.
2. Hutchison,N.J., Langer-Safer,P.R., Ward,D.C. and Hamkalo,B.A. (1982) *J. Cell Biol.*, **95**, 609.
3. Langer-Safer,P.R., Levine,M. and Ward,D.C. (1982) *Proc. Natl. Acad. Sci. USA*, **79**, 4381.
4. Cox,K.H., DeLeon,D.V., Angerer,L.M. and Angerer,R.C. (1984) *Dev. Biol.*, **101**, 485.
5. Maniatis,T., Jeffrey,A., van deSande,H. (1975) *Biochemistry (Wash.)*, **14**, 3787.
6. Gall,J.G. and Pardue,M.L. (1971) in *Methods in Enzymology*, Vol. **21**, Grossman,L. and Moldave,K. (eds.), Academic Press, New York, p. 470.
7. Melton,D.A., Krieg,P.A., Rebagliati,M.R., Maniatis,T., Zinn,K. and Green,M.R. (1984) *Nucleic Acids Res.*, **12**, 7035.
8. Zinn,K., Di Maio,D. and Maniatis,T. (1983) *Cell.* **34**, 865.
9. Maniatis,T., Jeffrey,A. and Kleid,D.G. (1975) *Proc. Natl. Acad. Sci. USA*, **72**, 1184.
10. Rigby,P.W.J., Dieckman,M., Rhodes,C. and Berg,P. (1977) *J. Mol. Biol.*, **113**, 237.
11. Yunis,J.J., Sawyer,J.R. and Ball,D.W. (1978) *Cytogenet. Cell Genet.*, **22**, 679.
12. Bonner,J.J. and Pardue,M.L. (1976) *Chromosoma*, **58**, 87.
13. Pardue,M.L., Gerbi,S.A., Eckhardt,R.A. and Gall,J.G. (1970) *Chromosoma*, **29**, 268.
14. Hayashi,S., Gillham,I.C., Delaney,S.D. and Tener,G.M. (1978) *J. Histochem. Cytochem.*, **26**, 677.
15. Wetmur,J.G. (1975) *Biopolymers*, **14**, 2517.
16. Wahl,G.M., Stern,M. and Stark,G.R. (1979) *Proc. Natl. Acad. Sci. USA*, **76**, 3683.
17. Harper,M.E. and Saunders,G.F. (1981) *Chromosoma*, **83**, 431.
18. Harper,M.E., Ullrich,A. and Saunders,G.F. (1981) *Proc. Natl. Acad. Sci. USA*, **78**, 4458.
19. Angerer,L.M. and Angerer,R.C. (1981) *Nucleic Acids Res.*, **9**, 2819.
20. Jamrich,M., Mahon,K., Gravis,E. and Gall,J.G. (1984) *EMBO J.*, **3**, 1939.
21. Hafen,E., Levine,M., Garber,R.L. and Gehring,W.J. (1983) *EMBO J.*, **2**, 617.
22. Akam,M. (1983) *EMBO J.*, **2**, 2075.
23. Wetmur,J.G., Ruyechan,W.T. and Douthart,R.T. (1981) *Biochemistry (Wash.)*, **20**, 2999.
24. Gray,D.M., Liu,J.-J., Ratliff,R.I. and Allen,F.S. (1981) *Biopolymers*, **20**, 1337.
25. Chandler,M.E. and Yunis,J.J. (1978) *Cytogenet. Cell Genet.*, **22**, 352.

APPENDIX I

Restriction Enzymes

R.J.ROBERTS

Restriction enzymes are endodeoxyribonucleases that recognise short, specific sequences within DNA molecules and then catalyse double-strand cleavage of the DNA. Three distinct classes of restriction enzymes are known:

Type I enzymes recognise a specific sequence in DNA and cleave the DNA chain at random locations with respect to that sequence. They have an absolute requirement for the co-factors S-adenosylmethionine and ATP, and during cleavage they hydrolyse the ATP. Because of the random nature of the cleavage, the products are a heterogeneous array of DNA fragments.

Type II enzymes also recognise a specific nucleotide sequence, but differ from the Type I enzymes in cleaving only at a specific location within or close to the recognition sequence, thus generating a unique set of fragments. At the present time, more than 400 such enzymes have been characterised, among which more than 100 different sequence specificities occur. In cases where enzymes from different sources recognise the same sequence, then the enzymes are called isoschizomers.

Type III enzymes have properties intermediate between those of the Type I and the Type II enzymes. They recognise a specific DNA sequence and cleave a short distance away from it. However, it is usually difficult to obtain complete digestion. They have an absolute requirement for ATP but they do not hydrolyse it, and although their activity is stimulated by S-adenosylmethionine it is not an absolute requirement.

In this Appendix, restriction enzymes and their isoschizomers are listed alphabetically by prototype. Their availability from three major commercial sources is indicated, as are the buffer conditions recommended by the manufacturer. It should be noted that for most restriction enzymes, their activity varies little over a wide range of ionic strength and pH, and the values listed in general have not rigorously been shown to be optimal. The information in this list is taken from Roberts,R.J., *Nucleic Acids Res.* (1984), **12,** r167-r204, and the New England Biolabs catalogue (1984 edition).

Enzyme	Isoschizomers	Recognition sequence	Commercial source[a]	Digestion conditions[b]					
				NaCl[d]	Tris	pH	Mg	SH	Other[e]
AatII		GACGT↑C	NEB	50 (K)	10	7.5	6	6	
AccI		GT↑(A/C)(G/T)AC	NEB,BRL,B-M	6	6	7.5	6	6	
AcyI		GPu↑CGPyC							
	AhaII		NEB	100	10	8.0	10	6	
	AosII								
	AstWI								
	AsuIII								
	BbiI								
	HgiDI								
	HgiGI								
	HgiHII								
	NlaSII								
AflII		C↑TTAAG							
AflIII		A↑CPuPyGT							
AhaIII		TTT↑AAA	NEB,B-M	75	10	8.0	10	5	
	DraI		NEB	10	10	8.0	10	0	
AluI		AG↑CT	NEB,BRL,B-M	50	6	7.6	6	6	
	OxaI								
ApaI		GGGCC↑C	NEB,B-M	6	6	7.4	6	6	
AsuI		G↑GNCC							
	Cfr13I								
	Cfr4I								
	Cfr8I								
	Eco39I								
	Eco47II								
	NmuEII								
	NmuSI								
	Nsp(7524)IV								
	PspI								
	Sau96I		NEB,BRL,B-M	60	6	7.4	15	6	
	SdyI								
AsuII		TT↑CGAA							
	FspII								
	MlaI								
	Nsp(7524)V								
	NspBI								
AvaI		C↑PyCGPuG	NEB,BRL,B-M	60	6	8.0	10	6	
	AquI								
	AvrI								
	Nsp(7524)III								
AvaII		G↑G(A/T)CC	NEB,BRL,B-M	60	6	8.0	10	6	
	AflI								
	BamNx								
	CauI								
	ClmII								
	Eco47I								
	FdiI								
	HgiBI								
	HgiCII								
	HgiEI								
	HgiHIII								
	NspHII								
	SinI								
AvaIII		ATGCAT							
	NsiI	ATGCA↑T	NEB	150	6	8.4	6	0	
AvrII		C↑CTAGG							
BalI		TGG↑CCA	NEB,BRL	0	10	7.6	12	10	

Enzyme	Isoschizomers	Recognition sequence	Commercial source[a]	Digestion conditions[b]					
				NaCl[c]	Tris	pH	Mg	SH	Other[d]
*Bam*HI		G↓GATCC	NEB,BRL,B-M	150	6	7.9	6	0	
	*Aac*I								
	*Aae*I								
	*Bam*FI								
	*Bam*KI								
	*Bam*NI								
	*Bst*I								
	*Dds*I								
	*Gdo*I								
	*Gin*I								
	*Gox*I								
	*Rhs*I								
*Bbv*I		GCAGC (8/12)	NEB	0	6	8.0	6	6	
*Bcl*I		T↓GATCA	NEB,BRL,B-M	75 (K)	6	7.4	10	1	(DTT) 50°C
	*Atu*CI								
	*Bst*GI								
	*Cpe*I								
	*Sst*IV								
*Bgl*I		GCCNNNN↓NGGC	NEB,BRL,B-M	66	10	7.4	10	1	
*Bgl*II		A↓GATCT	NEB,BRL,B-M	100	10	7.4	10	10	
*Bin*I		GGATC (4/5)							
	*Bth*II								
*Bse*PI		GCGCGC							
	*Bso*PI								
	*Bsr*HI								
	*Bss*HII	G↓CGCGC	NEB	25	6	7.4	6	10	50°C
*Bst*EII		G↓GTNACC	NEB,BRL,B-M	150	6	7.9	6	6	60°C
	*Asp*AI								
	*Bst*PI								
	*Cfr*7I								
	*Eca*I								
*Bst*XI		CCANNNNN↓NTGG	NEB	150	6	7.6	6	6	
	*Bss*GI								
	*Bst*TI								
*Cau*II		CC↓(G/C)GG							
	*Aha*I								
	*Bcn*I	CC(C/G)↓GG							
	*Nci*I		NEB,BRL,B-M	25	10	7.4	10	6	
*Cfr*10I		PuCCGGPy							
*Cfr*I		Py↓GGCCPu							
	Cfr14I								
	*Eae*I								
*Cla*I		AT↓CGAT	NEB,BRL,B-M	50	6	7.9	6	0	
	*Ban*III								
*Dde*I		C↓TNAG	NEB,BRL	150	6	7.5	6	6	
*Dpn*I		GA̅↓TC	NEB	150	10	7.4	10	10	
*Eco*47III		AGCGCT							
*Eco*A (Type 1)		GAG(N)7GTCA							
*Eco*B (Type 1)		TGA(N)8TGCT							
*Eco*DXI (Type 1)		ATCA(N)7ATTC							
*Eco*K (Type 1)		AAC(N)6GTGC							
*Eco*P15 (Type 3)		CAGCAG							
*Eco*PI (Type 3)		AGACC							
*Eco*RI		G↓AATTC	NEB,BRL,B-M	50	100	7.5	5	0	
	*Rsr*I								

Enzyme	Isoschizomers	Recognition sequence	Commercial source[a]	NaCl[c]	Tris	pH	Mg	SH	Other[d]
*Eco*RII		↑CC(A/T)GG	BRL	50	50	8.0	5	1	(DTT)
	*Aor*I	CC↑(A/T)GG							
	*Apy*I	CC↑(A/T)GG							
	*Atu*BI								
	*Atu*II								
	*Bin*SI								
	*Bst*GII								
	*Bst*NI	CC↑(A/T)GG	NEB	20 (K)	10	8.0	6	6	
	*Cfr*11I								
	*Cfr*5I								
	*Eca*II								
	*Ecl*II								
	*Eco*27I								
	*Eco*38I								
	*Mph*I								
	*Mva*I								
	*Taq*XI	CC↑(A/T)GG							
*Eco*RV		GAT↑ATC	NEB,BRL,B-M	150	6	7.9	6	6	
	*Eco*32I								
*Fnu*4HI		GC↑NGC	NEB	6	6	7.4	6	6	(5 mM KPO₄)
*Fnu*DII		CG↑CG	NEB	6	6	7.4	6	6	
	*Acc*II								
	*Bce*FI								
	*Bce*R								
	*Bsu*1192II								
	*Bsu*1193								
	*Bsu*EII								
	*Hin*1056I								
	*Tha*I		BRL	0	50	8.0	10	0	
*Fok*I		GGATG (9/13)	NEB	20 (K)	10	7.5	10	6	
	*Hin*GUII								
*Gdi*II		Py↑GGCCG							
*Gsu*I		CTCCAG							
	*Gsb*I								
*Hae*I		(A/T)GG↑CC(A/T)							
*Hae*II		PuGCGC↑Py	NEB,BRL,B-M	50	6	7.4	6	6	
	*Hin*HI								
	*Ngo*I								
*Hae*III		GG↑CC	NEB,BRL,B-M	50	6	7.4	6	6	
	*Blu*II								
	*Bse*I								
	*Bsp*RI								
	*Bss*CI								
	*Bst*CI								
	*Bsu*1076								
	*Bsu*1114								
	*Bsu*RI								
	*Clm*I								
	*Clt*I								
	*Fnu*DI								
	*Hhg*I								
	*Mni*I								
	*Mnn*II								
	*Ngo*II								
	*Nla*I								
	*Pai*I								
	*Pal*I								

Enzyme	Isoschizomers	Recognition sequence	Commercial source[a]	Digestion conditions[b]					
				NaCl[c]	Tris	pH	Mg	SH	Other[d]
	*Ppu*I								
	*Sfa*I								
	*Ttn*I								
	*Vha*I								
*Hga*I		GACGC (5/10)	NEB	50	6	7.4	6	6	
*Hgi*AI		G(A/T)GC(A/T)↑C	NEB	200	10	8.0	10	10	
*Hgi*CI		G↑GPyPuCC	NEB						
	*Ban*I		NEB	6	6	7.4	6	6	
	*Hgi*HI								
*Hgi*EII		ACC(N)6GGT							
*Hgi*JII		GPuGCPy↑C							
	*Ban*II		NEB	50	6	7.4	6	10	
	*Bvu*I								
	*Eco*24I								
	*Eco*25I								
	*Eco*26I								
	*Eco*35I								
	*Eco*40I								
	*Eco*41I								
*Hha*I		GCG↑C	NEB,BRL	0	50	8.0	5	0.5	(DTT)
	*Cfo*I		BRL	0	100	7.5	5	6	
	*Fnu*DIII								
	*Hin*GUI								
	*Hin*P1I	G↑CGC	NEB	0	50	8.0	5	0.5	(DTT)
	*Hin*S1								
	*Hin*S2								
	*Mnn*IV								
	*Sci*NI	G↑CGC							
*Hind*II		GTPy↑PuAC	B-M	50	10	7.6	10	14	(DTE)
	*Chu*II								
	*Hin*JCI								
	*Hinc*II		NEB,BRL	100	10	7.4	7	1	(DTT)
	*Mnn*I								
*Hind*III		A↑AGCTT	NEB,BRL,B-M	50	50	8.0	10	0	
	*Bbr*I								
	*Bpe*I								
	*Chu*I								
	*Hin*173								
	*Hin*JCII								
	*Hin*bIII								
	*Hinf*II								
	*Hsu*I								
	*Mki*I								
*Hinf*I		G↑ANTC	NEB,BRL	100	6	7.4	6	6	
	*Fnu*AI								
	*Hha*II								
	*Nca*I								
	*Nov*II								
	*Nsi*HI								
*Hinf*III (Type 3)		CGAAT							
	*Hine*I								
*Hpa*I		GTT↑AAC	NEB,BRL,B-M	20 (K)	10	7.4	10	1	(DTT)
	*Bse*II								
*Hpa*II		C↑CGG	NEB,BRL,B-M	6 (K)	10	7.4	10	1	(DTT)
	*Bsu*1192I								
	*Bsu*FI								
	*Hap*II								
	*Mni*II								

Enzyme	Isoschizomers	Recognition sequence	Commercial source[a]	Digestion conditions[b]					
				NaCl[c]	Tris	pH	Mg	SH	Other[d]
	*Mno*I								
	*Msp*I		NEB,BRL,B-M	6 (K)	10	7.4	10	1	(DTT)
	*Sfa*GU I								
*Hph*I		GGTGA (8/7)	NEB	6 (K)	10	7.4	10	1	(DTT)
*Kpn*I		GGTAC↓C	NEB,BRL,B-M	6	6	7.5	6	6	
	*Nmi*I								
*Mae*I		C↓TAG							
*Mae*II		A↓CGT							
*Mae*III		↓GTNAC							
*Mbo*I		↓GATC	NEB,BRL,B-M	100	10	7.4	10	1	(DTT)
	*Bsa*PI								
	*Bsr*PII								
	*Bss*GII								
	*Bst*EIII								
	*Bst*XII								
	*Cpa*I								
	*Cpf*I								
	*Dpn*II								
	*Fnu*AII								
	*Fnu*CI								
	*Fnu*EI								
	*Mno*III								
	*Mos*I								
	*Mth*I								
	*Nde*II		BRL	150	100	7.6	10	1	(DTT)
	*Nfl*I								
	*Nla*II								
	*Nsi*AI								
	*Nsu*I								
	*Pfa*I								
	*Sau*3A		NEB,BRL,B-M	50	6	7.5	5	0	
	*Sin*MI								
*Mbo*II		GAAGA (8/7)	NEB,BRL	6 (K)	10	7.4	10	1	(DTT)
	*Ncu*I								
*Mlu*I		A↓CGCGT	NEB	50	10	7.5	7	6	
*Mnl*I		CCTC (7/7)	NEB	150	12	7.6	12	6	
*Mst*I		TGC↓GCA	NEB	150	6	7.4	6	6	
	*Aos*I								
	*Fdi*II								
	*Fsp*I								
*Nae*I		GCC↓GGC	NEB	20	6	8.0	6	6	
	*Nba*I								
	*Nbr*I								
	*Nmu*I								
	*Pgl*I								
	*Rlu*I								
	*Ska*I								
*Nar*I		GG↓CGCC	NEB	0	6	7.4	6	6	
	*Bbe*AI								
	*Bbe*I	GGCGC↓C								
	*Bin*SII								
	*Nam*I								
	*Nda*I								
	*Nun*II								
*Nco*I		C↓CATGG	NEB,BRL	150	6	7.9	6	0	
*Nde*I		CA↓TATG	NEB,BRL	150	10	7.8	7	6	
*Nhe*I		GCTAGC							
*Nla*III		CATG↓	NEB	50 (A)	10	7.6	10	10	
*Nla*IV		GGN↓NCC	NEB	50 (A)	10	7.4	10	10	

Enzyme	Isoschizomers	Recognition sequence	Commercial source[a]	Digestion conditions[b]					
				NaCl[c]	Tris	pH	Mg	SH	Other[d]
*Not*I		GC↑GGCCGC	NEB	150	6	7.9	6	6	
*Nru*I		TCG↑CGA	NEB,BRL	100	6	7.4	6	5	
	*Ama*I								
Nsp(7524)I		PuCATG↑Py							
	*Nsp*HI								
*Nsp*BII		C(A/C)G↑C(T/G)G							
*Pst*I		CTGCA↑G	NEB,BRL,B-M	100	10	7.5	10	0	
	*Bbi*I								
	*Bce*170								
	*Bsu*1247								
	*Cau*III								
	*Eae*PI								
	*Eco*36I								
	*Eco*481								
	*Eco*491								
	*Mau*I								
	*Noc*I								
	*Pma*I								
	*Sal*PI								
	*Sfl*I								
	*Ska*II								
	*Xma*II								
	*Xor*I								
*Pvu*I		CGAT↑CG	NEB,BRL,B-M	150	6	7.4	6	6	
	*Nbl*I								
	*Rsh*I								
	*Rsp*I								
	*Xni*I								
	*Xor*II		BRL	0	6	7.4	12	6	
*Pvu*II		CAG↑CTG	NEB,BRL,B-M	60	6	7.5	6	6	
	*Cfr*61								
*Rsa*I		GT↑AC	NEB,B-M	50	6	8.0	12	6	
*Sac*I		GAGCT↑C	NEB,B-M	0	6	7.4	6	6	
	*Eco*ICRI								
	*Sst*I		BRL	50	50	8.0	10	0	
*Sac*II		CCGC↑GG	NEB	0	6	7.5	6	6	
	*Bac*I								
	*Csc*I								
	*Ecc*I								
	*Mra*I								
	*Ngi*III								
	*Nla*SI								
	*Sbo*I								
	*Sfr*I								
	*Shy*I								
	*Sst*II		BRL	50	50	8.0	10	0	
	*Tgl*I								
*Sal*I		G↑TCGAC	NEB,BRL,B-M	150	6	7.9	6	6	
	*Hgi*CIII								
	*Hgi*DII								
	*Nop*I								
	*Rhe*I								
	*Rhp*I								
	*Rrh*I								
	*Rro*I								
	*Xam*I								
*Sau*I		CC↑TNAGG	B-M	75	10	7.2	10	0	
	*Cvn*I								
	*Mst*II		NEB	150	6	7.5	6	6	

Appendix I

Enzyme	Isoschizomers	Recognition sequence	Commercial source[a]	NaCl[c]	Tris	pH	Mg	SH	Other[d]
ScaI		AGT↑ACT	NEB	100	6	7.4	6	6	
ScrFI		CCNGG	NEB	50	6	7.6	6	6	
SduI		G(G/A/T)GC(C/A/T)C							
	Bsp1286	G(G/A/T)GC(C/A/T)↑C							
	Nsp(7524)II	G(G/A/T)GC(C/A/T)↑C							
SfaNI		GCATC (5/9)	NEB	150	10	7.5	10	0	
SfiI		GGCCNNNN↑NGGCC	NEB	50	10	7.8	10	10	
SmaI		CCC↑GGG	NEB,BRL,B-M	20 (K)	6	8.0	6	6	
	Cfr9I	C↑CCGGG							
	XmaI	C↑CCGGG	NEB	25	6	7.5	6	6	
SnaBI		TACGT↑A	NEB	50	10	7.4	10	10	
SnaI		GTATAC							
SpeI		ACTAGT							
SphI		GCATG↑C	NEB,BRL,B-M	150	6	7.4	6	10	
	SpaI								
SspI		AATAATT							
StuI		AGG↑CCT	NEB,B-M	100	10	8.0	10	6	
	AatI								
	GdiI								
TaqI		T↑CGA	NEB,BRL,B-M	100	10	8.4	6	6	
	TfiI								
	TthHB8 I								
TaqII		GACCGA (11/9)							
		CACCCA (11/9)							
Tth111 I		GACN↑NNGTC	NEB	50	8	7.4	8	8	
	TteI								
	TtrI								
Tth111 II		CAAPuCA							
XbaI		T↑CTAGA	NEB,BRL,B-M	50	6	7.9	6	0	
XhoI		C↑TCGAG	NEB,BRL,B-M	150	6	7.9	6	6	
	BbiIII								
	BluI								
	BssHI								
	BstHI								
	BsuM								
	BthI								
	CcrII								
	DdeII								
	MsiI								
	PaeR7								
	PanI								
	ScuI								
	SexI								
	SgaI								
	SgoI								
	SlaI								
	SluI								
	SpaI								
	XpaI								
XhoII		Pu↑GATCPy	NEB,B-M	0	10	8.0	10	0	
XmaIII		C↑GGCCG	NEB,BRL	25	6	8.2	6	6	
XmnI		GAANN↑NNTTC	NEB	50	10	8.0	10	6	

[a]The abbreviations used are NEB (New England Biolabs), BRL (Bethesda Research Laboratories) and B-M (Boehringer-Mannheim). Several other sources of supply for these enzymes also exist.
[b]The concentrations of reagents are given in mM and are those recommended by the first manufacturer listed. Tris-HCl, MgCl$_2$ and 2-mercaptoethanol (SH) are used unless otherwise indicated. All digests are run at 37°C unless otherwise indicated and include 100 μg/ml bovine serum albumin.
[c](K) or (A) indicates that NaCl is replaced by KCl or (NH$_4$)$_2$SO$_4$, respectively.
[d]DTT (dithiothreitol) or DTE (dithioerythritol) indicate replacements for 2-mercaptoethanol whereas 50°C or 60°C indicate the recommended temperature for digestion where this is not 37°C.

210

Nucleic Acid Size Markers[1]

S.J. MINTER, P.G. SEALEY and J.E. ARRAND

1. DNA SIZE MARKERS

The following sections give sizes of the restriction fragments generated from commonly-used plasmid and bacteriophage DNAs by a variety of restriction endonucleases. These serve as useful size markers during gel electrophoresis of nucleic acid samples when detection is by staining with ethidium bromide. Alternatively, the restriction fragments can be radiolabelled by the following methods:

(i)   *Fragments with 5′ extensions.* The 5′ ends can be labelled by reaction with T4 polynucleotide kinase (either by exchange reaction or following dephosphorylation) using [γ-^{32}P]ATP. Alternatively, the 3′ ends can be labelled by 'filling in' opposite the 5′ extensions using the Klenow fragment of *Escherichia coli* DNA polymerase I plus the appropriate [α-^{32}P]deoxynucleotide triphosphate(s).

(ii)   *Fragments with 3′ extensions.* The 3′ ends are most easily labelled using terminal deoxynucleotidyltransferase (TdT) and [α-^{32}P]cordycepin triphosphate.

(iii)   *Blunt-ended fragments.* Either the 5′ ends can be labelled with kinase reactions as described in (i) above or the 3′ ends can be labelled using terminal deoxynucleotidyltransferase as described in (ii).

Detailed protocols for each of these radiolabelling methods are given in Chapter 2.

When determining the size of unknown DNA fragments, it should be noted that the reciprocal plot method of analysis (2) is more accurate in describing the mobility/size relationship, over a wider size range, than the conventional semi-log plot (3). This does not apply to RNA.

1.1 Plasmid pBR322

Plasmid pBR322 is a double-stranded circular DNA molecule. It is extremely convenient for use as a molecular weight marker since a wide range of fragment sizes can be generated by using a relatively small range of restriction endonucleases. The linear 4362 bp molecule (4), useful as a marker for agarose gels, can be obtained using any one of the following restriction enzymes, each of which cleaves at a single site in the plasmid:

*Aat*II, *Afl*III, *Ava*I, *Bal*I, *Bam*HI, *Cla*I, *Eco*RI, *Eco*RV, *Hind*III, *Hinc*I, *Nde*I, *Nru*I, *Pst*I, *Pvu*I, *Pvu*II, *Rru*I, *Sal*I, *Sca*I, *Sna*I, *Sph*I, *Tth*111I, *Xma*III, *Xor*II.

Other enzymes which have multiple restriction sites, together with the sizes of the fragments obtained, are listed in *Table 1*.

---

[1]This appendix is updated and extended from Appendix I of ref. 1.

Table 1. Sizes of Restriction Fragments of pBR322.[a,b,c]

BvuI	HincII	AccI	XmnI	EcoK	AhaIII	BglI	RsaI	NarI	NaeI	ThnII
4348	3256	2767	2430	2377	3651	2319	2117	3571	3481	3196
14	1106	1595	1932	1985	692	1809	1565	657	367	1127
					19	234	680	113	354	32
								21	160	7

MstI	HgiDI	BstNI	TaqI	AvaII	DdeI	HgiAI	HgiCI	HinfI	HaeII	HgaI
2134	2699	1857	1444	1746	1652	1161	2027	1631	1876	867
1095	657	1060	1307	1433	542	826	1123	517	622	731
1035	490	928	475	303	540	604	439	506	439	633
98	382	383	368	279	465	587	294	396	430	578
	113	121	315	249	426	498	218	344	370	415
	21	13	312	222	409	310	113	298	227	314
			141	88	166	291	84	221	181	245
				42	162	85	43	220	83	239
							21	154	60	158
								75	53	150
									21	32

NciI	MboII	FokI	HphI	Sau96I	AluI	ScrFI	Sau3A	SfaNI	HaeIII	TacI
724	790	1171	1106	1461	910	696	1374	1052	587	581
699	755	853	853	616	659	592	665	424	540	493
696	753	659	576	352	655	525	358	395	504	452
632	592	649	415	279	521	363	341	375	458	372
363	492	287	387	274	403	351	317	248	434	355
351	271	188	282	249	281	347	272	243	267	341
328	254	181	227	222	257	328	258	234	234	332
308	196	141	221	191	226	308	207	223	213	330
226	109	78	207	189	136	218	105	220	192	145
35	78	62	45	179	100	199	91	192	184	129
	72	48	34	124	63	184	78	164	124	129
		45	9	88	57	121	75	134	123	122
				79	49	42	46	96	104	115
				42	19	40	36	88	89	103
				17	15	35	31	78	80	97
					11	13	27	63	64	69
							18	39	57	66
							17	37	51	61
							15	25	21	27
							12	12	18	26
							11	11	11	10
							8	9	7	5
										2

[a]The enzymes are listed in order of increasing number of fragments generated.
[b]See page 211 for radiolabelling procedures for restriction fragments.
[c]Data derived using the pBR322 sequence on file in the EMBL DNA Sequence Library using the MAPSORT program (5).

Table 1 continued.

HpaII	MnlI	HhaI	Fnu4H
622	591	393	480
527	400	348	328
404	334	337	299
309	262	332	277
242	247	270	229
238	218	259	206
217	206	206	189
201	206	190	165
190	206	174	155
180	204	153	150
160	185	152	143
160	179	151	129
147	166	141	123
147	156	132	119
122	150	131	118
110	96	109	116
90	88	103	107
76	81	100	97
67	77	93	95
34	61	83	85
34	60	75	83
26	58	67	81
26	38	62	79
15	36	60	71
9	30	53	65
9	27	40	57
		36	53
		33	51
		30	46
		28	45
		21	34
			27
			18
			14
			7
			3×7

1.2 Plasmid pAT153

pBR322 and pAT153 are closely-related plasmids. pAT153 was derived from pBR322 by removing the B (622 bp) and G (83 bp) *Hae*II restriction fragments of pBR322 (ref. 6). Linear molecules of pAT153 are 3657 bp long; they can be obtained by restriction with any of the following enzymes:

*Aat*II, *Acc*I, *Ava*I, *Bal*I, *Bam*HI, *Cla*I, *Eco*K, *Eco*RV, *Hind*III, *Nru*I, *Pst*I, *Pvu*I, *Rru*I, *Sal*I, *Sph*I, *Xma*III, *Xmn*I, *Xor*II.

Other enzymes which have multiple restriction sites, together with the sizes of the fragments obtained, are listed in *Table 2*.

Table 2. Sizes of Restriction Fragments of pAT153[a].

*Bvu*I	*Rsa*I	*Hinc*II	*Tth*II	*Aha*III	*Bgl*I	*Nar*I	*Nae*I	*Mst*I	*Hgi*DI	*Bst*NI	*Dde*I
3643	3486	2551	3618	2946	1809	2866	2776	1429	1994	1857	1652
14	165	1106	32	692	1614	657	367	1095	657	928	540
			7	19	234	113	354	1035	490	383	464
						21	160	98	382	355	426
									113	121	409
									21	13	166

*Fok*I	*Taq*I	*Ava*II	*Hgi*AI	*Nci*I	*Hae*II	*Hinf*I	*Hgi*CI	*Hga*I	*Hph*I	*Mbo*II	*Alu*I
1584	1444	1433	1161	724	1876	1631	1322	867	986	790	659
853	602	1320	619	696	439	517	1117	731	853	755	655
659	475	303	604	665	430	396	439	578	415	592	622
287	368	249	587	632	370	298	294	501	387	492	521
181	315	222	310	363	227	221	218	314	282	271	403
48	312	88	291	351	181	220	113	245	227	254	257
45	141	42	85	226	60	154	84	239	221	196	226
					53	145	43	150	207	109	136
					21	75	21	32	45	78	100
									34	72	63
										48	15

[a]See footnotes a and b of *Table 1*.

Table 2 continued.

Sau96	SfaNI	ScrFI	TacI	Sau3A	MnlI	HaeIII	HpaII	HhaI	Fnu4H
1224	1052	696	581	876	591	587	622	393	480
616	437	592	493	665	400	458	492	337	328
352	424	525	452	358	247	434	404	332	277
274	395	363	355	341	218	339	242	270	229
249	375	351	332	317	206	267	238	259	206
222	248	313	330	272	206	234	217	206	189
191	234	218	182	258	206	213	201	190	165
179	192	199	145	105	204	192	190	174	155
124	164	184	129	91	200	184	160	153	150
88	88	121	129	78	179	124	160	152	143
79	25	42	122	75	166	123	147	151	129
42	12	40	115	46	156	104	122	132	119
17	11	13	97	36	150	89	110	131	118
			66	31	96	80	90	109	113
			61	27	88	64	76	100	107
			27	18	81	57	67	93	95
			26	17	77	51	34	75	85
			10	15	61	21	26	67	83
			5	12	60	18	26	62	79
				11	38	11	15	60	71
				8	27	7	9	53	65
							9	40	57
								36	51
								33	45
								28	34
								21	27
									18
									14
									7
									3 x 6

1.3 **Bacteriophage lambda** (λ)

Bacteriophage λ is the most extensively-studied phage of *Escherichia coli*. The DNA of strain λ cI *indl ts* 857 Sam7 has been sequenced (7). The native molecule is circular, double-stranded and 48 502 bp in length. Heating at 68°C for 10 min, in a buffer containing a low concentration of salt, followed by rapid cooling causes separation of the cohesive ends at the 'cos' site. This generates a linear molecule with single-stranded 'sticky' ends 12 nucleotides long. *Table 3* lists the sizes of fragments generated by a variety of restriction endonucleases.

Table 3. Sizes of Restriction Fragments of Phage λ DNA (strain cI *ts* 857)[a,b,c].

*Nar*I	*Xho*I	*Xba*I	*Sal*I	*Kpn*I	*Cvn*I	*Sst*I	*Avr*II	*Xma*III	*Afl*II
45 680*	33 498*	24 508*	32 745*	29 942*	26 718*	24 776*	24 322*	19 944*	30 012
2822*	15 004*	23 994*	15 258*	17 057*	14 183*	22 621*	24 106*	16 710	6540*
			499	1503	7601	1105	74	11 848*	6078*
									5872

*Sma*I	*Pvu*I	*Sst*II	*Nco*I	*Nru*I	*Eco*RI	*Bam*HI	*Bgl*II	*Sph*I	*Nde*I
19 399*	14 321	20 323*	19 329*	23 460	21 226*	16 841	22 010	12 044	27 631*
12 220	12 712*	18 780	16 380	9401	7421	7233	13 286	11 940	8370*
8612*	11 936*	8113*	4572	6692*	5804	6770*	9688*	9790	3796
8271	9533	1076	4254*	4592*	5643	6527	2392	8090*	2433
		210	3967	3653	4878	5626	651	3003	2253
				704	3530*	5505*	415*	2216*	1774
							60	429	1689
									556

*Hind*III	*Bcl*I	*Ava*I	*Acc*I	*Hpa*I	*Pvu*II
23 130*	18 909	14 677	13 070	8666*	21 088*
9416	8844*	8614*	11 828	6911	4421
6557	6330	6888	6957	5414	4268
4361*	4623	4720*	5580*	4535	4194
2322	4459	4716	3574	4491	3916
2027	2684	3730	2720	4347	3638
564	1576	1881	2191*	3408	2296
125	560*	1674	1444	3384	1708
	517	1602	639	3042	636
			499	2240	579
				734*	532
				441	468
				410	343
				251	211*
				228	141
					63

[a]See footnotes a and b of *Table 1*.
[b]Data derived from the sequence of λ cI *ts* 857 on file in the EMBL DNA Sequence Library using the MAPSORT program (5). Fragment sizes are calculated to the cut site of the recognition sequence (rather than to the 5' nucleotide of the recognition sequence as in some commercial catalogues).
[c]Note that some λ fragments cross-hybridise with sequences contained in plasmid pBR322 and its relatives (pAT153, pUC8 and 9, etc). This may cause problems if Southern blots containing λ markers are probed with sequences cloned into these plasmid vectors, and it may be advisable to cut away the part of the blot containing the markers prior to autoradiography to avoid blurring.
* indicates which λ fragments have cos ends. These may reassociate when the markers are stored, generating new bands in gels. To avoid this problem the markers should be heated prior to use as described in Section 1.3 text.

1.4 **Simian Virus 40 (SV40)**

The SV40 genome is a double-stranded circular DNA, 5243 bp in size. It can be linearised using *Acc*I, *Bam*HI, *Bcl*I, *Bgl*I, *Eco*RI, *Eco*RV, *Hae*II, *Kpn*I, *Msp*I (*Hpa*II), *Nae*I or *Taq*I. *Table 4* lists the sizes of restriction fragments generated by other enzymes which have multiple cleavage sites.

Table 4. Sizes of Restriction Fragments from SV40[a,b,c].

*Hha*I	*Nde*I	*Pst*I	*Xho*II	*Pvu*II	*Hph*I	*Hpa*I	*Ava*II	*Hind*II	*Hinc*II
4753	4225	4027	3007	2007	3091	2147	1580	1768	1980
490	1018	1216	1566	1719	1856	2009	1525	1169	1538
		760	1446		160	1067	995	1116	1067
				136		20	682	526	369
							430	447	240
							31	215	29
									20

*Mbo*II	*Hinf*I	*Hae*I	*Aha*III	*Rsa*I	*Mbo*II	*Hae*III	*Alu*I	
1347	1845	1739	1753	1605	1350	1661	775	46
1264	1085	1661	739	708	756	765	483	41
945	766	383	565	675	687	540	329	38
610	543	348	430	551	645	373	288	30
396	525	329	411	497	409	329	275	29
384	237	300	364	351	395	325	253	28
234	109	227	318	294	383	300	253	27
60	83	179	315	226	375	299	243	12
	24	33	141	153	69	227	224	10
	24	30	136	111	65	179	223	8
		14	71	57	31	48	177	7
				15	30	45	157	
					14	41	154	
					13	33	153	
					11	30	146	
					10	29	144	
						14	123	
						9	75	
						6	54	
							53	
							50	
							49	

[a]See footnotes a and b of *Table 1*.
[b]From sequence data for SV40 DNA reported in references 8 – 10.
[c]There are no cleavage sites for *Ava*I, *Bal*I, *Bgl*II, *Bpa*I, *Cla*I, *Fnu*DII, *Pvu*I, *Sal*I, *Sma*I, *Sst*I, *Sst*II, *Sst*III, *Xba*I, *Xho*I or *Xma*II.

1.5 Coliphage M13 mp7

This is a widely-used cloning and sequencing vector which consists of a circular DNA molecule, 7238 bp long (RF form). It can be linearised by *Acy*I, *Ava*I, *Ava*II, *Avr*I, *Bgl*I, *Bgl*II, *Mst*II, *Nae*I, *Nar*I, *Pvu*I, *Pst*I, *Sau*I, *Sua*I. *Table 5* lists the sizes of fragments generated from M13 mp7 DNA by enzymes which have multiple restriction sites.

Table 5. Sizes of Restriction Fragments for Phage M13 mp7[a,b].

AccI	SalI	HincII	BamHI	EcoRI	GdiII	PvuII	HgiAI	XmnI	ClaI
7226	7226	7226	7214	7196	6993	6835	6516	4949	4343
12	12	12	24	42	245	310	722	2289	2895

AsuI	CauII	NciI	XhoI	HaeI	HgiCI	HaeII	EcoRII	SfaNI	HgaI
6578	4313	4313	4020	2836	4428	3514	3975	2625	1638
446	4162	2336	2535	2813	2022	2520	1809	1684	1517
190	558	558	659	1325	324	434	952	966	1089
24	31	31	24	264	322	433	179	871	1075
					130	329	139	716	846
					12	8	127	363	758
							57	13	315

Sau3A	BbvI	KpnI	TaqI	HaeIII	Fnu4HI	HphI	RsaI	FnuDII	HpaII
4020	1826	1694	1018	2527	1826	2212	1345	1290	1596
1696	1739	1241	971	1623	1739	1595	1334	889	829
507	1154	1226	927	849	891	1013	1004	772	818
434	665	792	703	341	629	589	742	681	651
332	611	666	639	311	611	489	604	646	545
129	436	332	612	309	436	271	522	541	543
96	420	318	579	245	420	241	322	496	472
24	387	301	564	214	387	194	258	495	454
		196	441	169	164	144	201	361	357
		196	381	158	45	135	190	353	183
		163	239	117	27	127	163	304	176
		113	152	106	24	76	143	190	156
			12	102	22	39	107	63	130
				98	14	39	102	57	123
				69	3	33	93	54	79
						26	65	24	60
						9	27	20	30
						6	16	2	18
									18

[a]See footnotes a and b of *Table 1*.
[b]There are no cleavage sites for *Bst*EII, *Hind*III, *Hpa*I, *Kpn*I, *Mst*I, *Sma*I, *Xho*I or *Xma*I.

Table 5 continued.

HhaI	AluI	HinfI	DdeI	EcoRI*	MnlI	
967	1446	1288	998	1402	423	31
732	1330	771	866	665	421	30
725	600	571	672	567	391	28
714	555	486	652	508	384	27
683	484	413	600	416	351	22
495	336	348	563	406	340	21
434	331	345	399	323	318	19
340	313	328	381	283	313	18
329	220	324	376	272	304	16
312	204	274	303	247	286	15
293	201	261	294	213	269	15
272	180	253	272	209	263	15
244	159	234	160	176	243	15
190	151	232	153	174	234	15
115	140	212	128	152	226	15
92	111	209	72	142	217	15
74	111	160	63	109	158	15
65	104	137	46	109	143	12
56	93	96	42	102	130	8
28	72	80	39	99	119	6
26	63	63	28	96	117	5
22	39	46	27	88	111	4
13	39	45	24	76	105	4
9	27	22	15	69	102	
8	26	21	15	63	98	
	24	19	15	55	90	
			15	53	86	
			15	41	76	
			14	40	74	
				36	69	
				33	66	
				12	63	
				1	51	
				1	49	
					48	
					48	
					43	
					38	

1.6 **Bacteriophage** ϕ**X174**

This is a bacteriophage of *E. coli* with a single-stranded circular DNA molecule. The RF form is double-stranded and 5386 bp long. It can be linearised using *Aos*I, *Aat*II, *Ava*I, *Ava*II, *Bse*PI, *Gdi*II, *Hgi*EII, *Mst*II, *Ncl*I, *Pst*I, *Sac*II, *Sst*II, *Stu*I or *Xho*I. The restriction fragment sizes for enzymes with multiple restriction sites are given in *Table 6*.

Table 6. Sizes of Restriction Fragments of Phage ϕX174[a,b,c].

*Aha*III	*Sau*96I	*Nar*I	*Nru*I	*Acc*I	*Eco*RII	*Xmn*I	*Hpa*I
4307	4064	3429	3222	3034	2767	4126	3730
1079	1322	1957	2164	2352	2619	974	1264
						286	392

*Hpa*II	*Hae*I	*Hae*II	*Hph*I	*Taq*I	*Rsa*I	*Hae*III	*Mbo*II
2748	2712	2314	1638	2914	1560	1353	1103
1697	872	1565	1116	1175	964	1078	1064
374	738	783	791	404	645	872	857
348	603	269	777	327	525	605	812
219	389	185	777	231	472	310	396
	72	125	110	141	392	281	396
		93	77	87	247	271	324
		54	43	54	197	234	224
			39	33	157	194	118
				20	138	118	89
					89	72	3

*Hinc*II	*Dde*I	*Fnu*DII	*Hha*I	*Hinf*I	*Alu*I		
1057	1012	1050	1553	726	1007	25	
770	998	870	640	713	853	21	
612	927	718	614	553	662	9	
495	542	695	532	500	358		
392	486	530	305	427	337		
345	393	496	300	417	276		
341	303	259	269	413	258		
335	302	170	201	311	254		
297	186	156	192	249	247		
291	165	127	145	200	204		
210	38	114	143	151	146		
162	18	103	123	140	119		
79	10	79	101	118	109		
	6	19	93	100	90		
			84	82	87		
			54	66	84		
			35	66	78		
			2	48	55		
				42	42		
				40	33		
				24	32		

[a]See footnotes a and b of *Table 1*.
[b]Data from reference 11.
[c]There are no cleavage sites for *Apa*I, *Asu*II, *Ava*III, *Avr*II, *Bal*I, *Bam*HI, *Bcl*I, *Bgl*I, *Bgl*II, *Bin*I, *Bst*EII, *Bst*XI, *Bvu*I, *Cla*I, *Cvn*I, *Eco*RI, *Eco*K, *Eco*RV, *Hind*III, *Kpn*I, *Mst*II, *Nae*I, *Nco*I, *Nde*I, *Nsp*HI, *Pvu*I, *Pvu*II, *Sal*I, *Sau*3A, *Sca*I, *Sma*I, *Sna*I, *Sph*I, *Sst*I, *Tth*111I, *Xba*I, *Xho*II, *Xma*I and *Xma*III.

1.7 **Coliphage fd**

This is a male-specific coliphage closely related to the phage M13. The virion is a single-stranded circular DNA molecule, 6408 nucleotides in size. The fd DNA can be linearised using either *Acc*I or *Hinc*II. The restriction fragment sizes for enzymes with multiple cleavage sites are given in *Table 7*.

Table 7. Sizes of Restriction Fragments for Phage fd DNA[a,b,c].

*Bam*HI	*Hae*II	*Mbo*II	*Hae*III	*Taq*I	*Msp*I(*Hpa*II)	*Alu*I
3425	3550	4349	2528	2019	1596	1446
2983	2033	666	1633	850	829	1330
	817	384	849	703	819	705
	8	332	352	652	652	554
		318	311	579	648	484
		196	309	441	501	366
		163	154	381	454	314
			106	357	381	257
			103	287	156	220
			69	139	129	204
					123	201
					60	166
					42	111
					12	29
					6	27
						24

[a]See footnotes a and b of *Table 1*.
[b]Data from reference 12.
[c]*Hinf*I gives 24 fragments and *Mnl*I gives 51 fragments. There are no cleavage sites for *Bcl*I, *Eco*RI, *Hin*dIII or *Kpn*I.

1.8 **Synthetic DNA Markers**

Synthetic DNA markers are now commercially available (e.g. BRL) and cover several ranges of fragment sizes. They circumvent the problems which occur with conventional markers since bands are evenly spaced throughout the gel, thus aiding the accuracy of size determination. Fragment sizes available in three such marker sets are shown in *Table 8*.

Table 8. Sizes of DNA Fragments in Synthetic Marker Sets[a].

1 kilobase ladder	123 base pair ladder	Oligo(dT) ladder
12 216	4182	22
11 198	4059	21
10 180	3936	20
9162	3813	19
8144	3690	18
7126	3567	17
6108	3444	16
5090	3321	15
4072	3198	14
3054	3075	13
2036	2952	12
1635	2829	11
1018	2706	10
516/506	2583	9
394	2460	8
344	2337	7
298	2214	6
211/210	2091	5
	1968	4
	1845	
	1722	
	1599	
	1476	
	1353	
	1230	
	1107	
	984	
	861	
	738	
	615	
	492	
	369	
	246	
	123	

[a]See footnote b of *Table 1*.

2. RNA SIZE MARKERS

The molecular weight of RNA can be determined by electrophoresis in non-denaturing agarose or polyacrylamide gels using RNA denatured prior to electrophoresis (e.g., using glyoxal or DMSO). Alternatively, one can use denaturing gels, either agarose containing methyl mercuric hydroxide or formamide or polyacrylamide gels containing formamide or 8 M urea (13). RNA migrates in proportion to the log of its molecular weight. In addition to the RNA molecular weight markers listed in *Table 9*, restriction endonuclease fragments of DNA (see *Tables 1 − 8*) may also be useful as size markers in certain situations.

Table 9. Molecular Weight Markers for Gel Electrophoresis of RNA.

RNA species[a]	*Molecular weight*[b]	*Number of nucleotides*	*Reference*
Myosin heavy chain mRNA (chicken)	2.02×10^6	6500	17
28S rRNA (HeLa cells)	1.90×10^6	6333	14
25S rRNA (*Aspergillus*)	1.24×10^6	4000	23
23S rRNA (*E. coli*)	1.07×10^6	3566	15
18S rRNA (HeLa cells)	0.71×10^6	2366	14
17S rRNA (*Aspergillus*)	0.62×10^6	2000	23
16S rRNA (*E. coli*)	0.53×10^6	1776	16
A2 crystallin mRNA (calf lens)	0.45×10^6	1460	18
Immunoglobulin light chain mRNA (mouse)	0.39×10^6	1250	19
β-globin mRNA (mouse)	0.24×10^6	783	20
β-globin mRNA (rabbit)	0.22×10^6	710	21
α-globin mRNA (mouse)	0.22×10^6	696	20
α-globin mRNA (rabbit)	0.20×10^6	630	21
Histone H4 mRNA (sea urchin)	0.13×10^6	410	22
5.8S RNA (*Aspergillus*)	4.89×10^4	158	23
5S RNA (*E. coli*)	3.72×10^4	120	24
4S RNA (*Aspergillus*)	2.63×10^4	85	23

[a]A more extensive list of sizes of rRNAs is given in ref. 25.
[b]Molecular weights are approximate only and based upon average 'molecular weight' of 310 for each nucleotide residue.

3. REFERENCES

1. Hames,B.D. and Higgins,S.J., eds. (1984) *Transcription and Translation − A Practical Approach,* IRL Press Ltd., Oxford and Washington, DC.
2. Southern,E.M. (1979) *Anal. Biochem.,* **100**, 319.
3. Sealey,P.G. and Southern,E.M. (1982) in *Gel Electrophoresis of Nucleic Acids − A Practical Approach,* Rickwood,D. and Hames,B.D. (eds.), IRL Press Ltd., Oxford and Washington DC, p.39.
4. Sutcliffe,J.G. (1979) *Cold Spring Harbor Symp. Quant. Biol.,* **43**, 77.
5. Devereux,J., Haeberli,P. and Smithies,O. (1984) *Nucleic Acids Res.,* **12**, 387.
6. Twigg,A.J. and Sherrat,D. (1980) *Nature,* **283**, 216.
7. Sanger,F., Coulson,A.R., Hong,G.F., Hill,D.F. and Petersen,G.B. (1982) *J. Mol. Biol.,* **162**, 729.
8. Fiers,W., Contreras,R., Haegeman,G., Rogiers,R., Van de Voorde,A., Van Heuverswyn,H., Van Herreweghe,J., Volckaert,G. and Ysebaert,M. (1978) *Nature,* **273**, 113.
9. Van Heuversuryn,H. and Fiers,W. (1979) *Eur. J. Biochem.,* **100**, 50.

10. Tooze,J., ed. (1980) *Molecular Biology of Tumor Viruses,* 2nd edition, Part 2, *DNA Tumor Viruses,* Cold Spring Harbor Laboratory Press, NY.
11. Sanger,F., Coulson,A.R., Friedmann,T., Air,G.M., Barrell,B.G., Brown,N.L., Fiddes,J.C., Hutchison,C.A., Slocombe,P.M. and Smith,M. (1978) *J. Mol. Biol.,* **125**, 225.
12. Beck,F., Sommer,R., Auerswald,E.A., Kurz,Ch., Zink,B., Osterburg,G. and Schaller,H. (1978) *Nucleic Acids Res.,* **5**, 4495.
13. Grierson,D. (1982) in *Gel Electrophoresis of Nucleic Acids — A Practical Approach,* Rickwood,D. and Hames,B.D. (eds.), IRL Press Ltd., Oxford and Washington, DC, p. 1.
14. McConkey,E. and Hopkins,J. (1969) *J. Mol. Biol.,* **39**, 545.
15. Stanley,W.M. and Bock,R.M. (1965) *Biochemistry (Wash.),* **4**, 1302.
16. Pearce,T.C., Rowe,A.J. and Turnock,G. (1975) *J. Mol. Biol.,* **97**, 193.
17. Mondal,H., Sutton,A., Chen,V.J. and Sarkar,S. (1974) *Biochem. Biophys. Res. Commun.,* **56**, 988.
18. Berns,A., Jansson,P. and Bloemendal,H. (1974) *Biochem. Biophys. Res. Commun.,* **59**, 1157.
19. Stavnezer,J., Huang,R.C.C., Stravnezer,E. and Bishop,J.M. (1974) *J. Mol. Biol.,* **88**, 43.
20. Morrison,M.R. and Lingrel,J.B. (1976) *Biochim. Biophys. Acta,* **447**, 104.
21. Hamlyn,P.H. and Gould,H.J. (1973) *J. Mol. Biol.,* **94**, 101.
22. Grunstein,M. and Schedl,P. (1976) *J. Mol. Biol.,* **104**, 323.
23. Scazzochio,C., personal communication.
24. Brownlee,G., Sanger,F. and Barrell,B.G. (1968) *J. Mol. Biol.,* **34**, 379.
25. Loening,U.E. (1968) *J. Mol. Biol.,* **38**, 355.

APPENDIX III

Computer Analysis of Nucleic Acid Hybridisation Data

B.D. YOUNG and M.L.M. ANDERSON

INTRODUCTION

DNA reassociation and RNA-DNA hybridisation reactions often consist of several kinetic components. If each component has a very different rate constant, it will be relatively easy to resolve the contribution made by each component. However, if the components are not clearly separated, resolution will be possible only by using computer techniques. The program described here uses a non-linear least squares technique to determine the components which best fit the experimental data. There are a number of techniques available for analysing data. The program given here is based on the most efficient algorithm which was originally developed by Marquadt (1) and described in detail by Bevington (2). This is an iterative technique in which the experimenter supplies initial values of the rate constants and fractions for each component, and the program converges to the least squares solution. Of all the techniques available, the Marquadt algorithm offers the most stable and rapid convergence. Similar programs have already been described by Kells and Strauss (3) and Pearson et al. (4).

DESCRIPTION OF THE PROGRAM

There are three possible functions (F) which the experimenter can select for the analysis of the data (*Table 1*). The number of components (m) into which the data are to be resolved is also selected in advance. Let the number of data points be n and the value of hybridisation or reassociation of the ith point be y_i. The value of the jth function at the ith $C_o t$ or $R_o t$ value is denoted F_{ij}. The background level, which is also variable, is denoted F_o. Hence the sum of the squares of the differences between the experimental data and the fitted curves is given by

$$Q = \sum_i^n (y_i - F_o - \sum_j^m F_{ij})^2$$

Table 1. Functions for Computer Analysis of Data[a]

$F = \dfrac{A}{1 + k\,C_o t}$	DNA reassociation by hydroxylapatite
$F = \dfrac{A}{(1 + k\,C_o t)^{0.45}}$	DNA reassociation by nuclease S1
$F = A(1 - e^{-k\,R_o t})$	RNA-DNA hybridisation by nuclease S1

[a]The choice of hydroxylapatite chromatography or resistance to digestion by nuclease S1 to monitor hybridisation is discussed in Chapter 3, Section 7.

The program is designed to find the values of the background F_o, and A and k for each component which produce the smallest value of Q. A further feature of this program is that any of these variables may be held constant if desired. The program can occasionally produce negative values of the parameters A and k. This is biologically meaningless and is usually an indication that either there is too much scatter on the data or that the experimenter is trying to resolve the data into too many components. For this reason we have not attempted to hold the parameters positive which might conceal some insufficiency in the data.

The standard error of each parameter is computed as described by Kells and Strauss (3). This is a useful feature since it indicates the uniqueness and reliability of the solution. If too many components are fitted or if poor data are analysed, the standard errors will often be $100-200\%$ of the parameter values. As pointed out by Pearson *et al.* (4), errors in the estimates of the parameters can be considerable even with moderately good data. Hence it is important that the experimenter should take this factor into account when drawing conclusions from the analysis of his experimental data.

PROGRAM LISTING

This program was written for the Wang 2200S computer. It is divided into three parts to allow operation on small microcomputers. Part I deals with the entry of data and initial parameters and includes routines for the display of data. Part II handles calculation of the least squares fit to the experimental data. Part III handles printing of the results and plotting of the fitted curve. The program listing is given below.

Part I

```
10COM A(13),A0(13),A1(13),A2(13),P(100,2),M,N,M0,D(13,2)
15COM T,L1,X1,X2,X4,Y1,Y2,Y4,B$64,S1,W,W0,L$(2)3,L9
17 L$(1)="COT":L$(2)="ROT"
18 SELECT PLOT 215
20 SELECT PRINT 005
30 PRINT HEX(03)
40PRINT "          NUCLEIC ACID REASSOCIATION PROGRAM"
50PRINT :PRINT :PRINT
60PRINT "THIS PROGRAM PLOTS AND ANALYSES ROT AND COT CURVES"
70PRINT :PRINT :PRINT
80IF W0=0THEN 190
90 INPUT "DO YOU WISH TO WORK ON CURRENT DATA",A$
100 IF A$="Y"THEN 275
190INPUT "DO YOU WISH TO INPUT EXPERIMENTAL DATA",A$
200 IF A$="Y"THEN 230
210 GOSUB 860
220 GOTO 275
230 INPUT "IS THIS A COT CURVE(1) OR A ROT CURVE(2)",L9
235 INPUT "NO OF POINTS",N:PRINTUSING 236,L$(L9)
236%INPUT DATA AS ###  ,%
238 PRINT "POINT NO"
240 FOR I=1TO N
244PRINTUSING 246,I;
246%###
250 INPUT P(I,1),P(I,2)
255 IF P(I,1))0THEN 270
260 PRINT L$(L9);" CANNOT=0"
265 GOTO 244
270NEXT I
275 PRINT HEX(03),"DATA POINTS=",N
280FOR I=1TO N
285PRINTUSING 290,P(I,1);P(I,2);
290%#.#↑↑↑↑,##.#    #.#↑↑↑↑,##.#    #.#↑↑↑↑,##.#
```

```
300NEXT I
305PRINT
310INPUT "DATA OK",A$:IF A$="Y"THEN 380
320 PRINT "INPUT POINT NUMBER,ROT,%"
330 INPUT I,P(I,1),P(I,2)
340 GOTO 275
380 INPUT "DO YOU WISH TO USE PLOTTER",A$
390 IF A$="N"THEN 400:GOSUB 1005
400PRINT HEX(03):PRINT "ANALYSIS OF DATA":PRINT :PRINT :GOSUB 17
00
405 INPUT "NO OF COMPONENTS TO BE FITTED",M
406INPUT "AUTOMATIC SELECTION OF STARTING VALUES",A$
407IF A$="N"THEN 410:GOSUB 2000:GOTO 670
410 INPUT "ALL PARAMETERS VARIABLE",A$
420 IF A$="N" THEN 510
430 PRINTUSING 435,L$(L9)
435 %INPUT ###1/2,%
440 FOR I=1 TO M
450 PRINT "COMPONENT",I
460 I0=2*(I-1)+1
470 INPUT A(I0),A(I0+1)
475 GOSUB 1750
480 A1(I0),A1(I0+1)=1
490 NEXT I
500 GOTO 590
510 PRINT "INPUT PARAMETERS,0=FIX,1=VAR."
520 FOR I=1 TO M
530 PRINT "COMPONENT",I
540 I0=2*(I-1)+1
550 INPUT "ROT1/2",A(I0),A1(I0)
555 GOSUB 1750
560 INPUT "%",A(I0+1),A1(I0+1)
580 NEXT I
590 INPUT "BACKGROUND ,0=FIXED,1=VARIABLE",A(I0+2),A1(I0+2)
670 M0=0:FOR I=1 TO M:FOR J=1 TO 2
700 I0=2*(I-1)+J
710 IF A1(I0)=0 THEN 740
720 M0=M0+1
730 D(M0,1)=J:D(M0,2)=I
740 NEXT J:NEXT I
750 IF A1(I0+1)=0THEN 820
760 M0=M0+1
770 D(M0,1)=1:D(M0,2)=M+1
820MAT A0=A
830 MAT A2=ZER
850LOAD "PART 2"
860 GOSUB 1700:L9=1:IF W<3THEN 863:L9=2
863 INPUT "NO OF POINTS",N
865 INPUT "NO OF COMPONENTS",M1:PRINTUSING 866,L$(L9)
866%####1/2,%
870 FOR I=1 TO M1:I0=2*(I-1)+1:INPUT A(I0),A(I0+1):GOSUB 1750
875 NEXT I
900 INPUT "BACKGROUND=",A(M1*2+1)
910 INPUT "LOWER LIMIT",R1:INPUT "UPPER LIMIT",R2
920 T=EXP((LOG(R2)-LOG(R1))/N)
930 FOR K=1 TO N:P(K,2)=A(M1*2+1)
940 P(K,1)=R1:FOR L=1 TO M1:GOSUB 980
950 P(K,2)=P(K,2)+F:NEXT L
960 R1=R1*T:PRINT P(K,1),P(K,2):NEXT K
970 MAT A=ZER:RETURN
980 ON W GOTO 990,995,1000
990 F=A(2*L)*(1-1/(1+P(K,1)*A(2*L-1))):RETURN
995 F=A(2*L)*(1-1/((1+P(K,1)*A(2*L-1))^0.44)):RETURN
1000 F=A(2*L)*(1-EXP(-P(K,1)*A(2*L-1))):RETURN
1005PRINT "SELECT PLOTTING MODE":PRINT "1.  STRAIGHT LINE"
1006PRINT "2.  SQUARES":PRINT "3.  TRIANGLES"
1007PRINT "4.  CROSSES":PRINT "5.  PLUS SIGNS"
1008INPUT R1:INPUT "AUTOMATIC SELECTION OF AXES",A$
1009IF A$<>"Y"THEN 1052
1010 Y2=0:X2=-1:X1=10000000
1020 FOR K=1 TO N:IF P(K,2)<Y2 THEN 1030:Y2=P(K,2)
1030 IF P(K,1)<X2THEN 1040:X2=P(K,1)
1040 IF P(K,1)>X1THEN 1045:X1=P(K,1)
1045 NEXT K
1050 Y1=0:Y2=Y2*1.1:IF Y2<100THEN 1060:Y2=100:GOTO 1060
1052INPUT "LOWER,UPPER LIMITS OF X-AXIS",X1,X2
```

```
1054INPUT "LOWER,UPPER LIMITS OF Y-AXIS",Y1,Y2
1060 X1=INT(LOG(X1)/LOG(10)):IF LOG(X2)/LOG(10)=INT(LOG(X2)/LOG(
10))THEN 1065:X2=INT(LOG(X2)/LOG(10))+1:GOTO 1070
1065X2=INT(LOG(X2)/LOG(10))
1070 Y3=4.5:X3=7
1072 INPUT "INVERTED PLOT",A$
1074 IF A$="N" THEN 1080:Y1=Y2:Y2=0 ·
1080Y4=Y3*100/(Y2-Y1):X4=X3*100/(X2-X1)
1100 PLOT <,,HEX(0603E5E62035E4))
1110GOSUB 1622
1130 P9=0:S=1
1140FOR K=1TO N:X=LOG(P(K,1))/LOG(10):Y=P(K,2):GOSUB 1200
1150 IF R1<>1THEN 1160:P9=1:GOTO 1165
1160 ON (R1-1)GOTO 1161,1162,1163,1164
1161PLOT <4*S,4*S,U>,<,-8*S,,D>,<-8*S,,D>,<,8*S,D>,<8*S,,D>,<-4*S
,-4*S,U>:GOTO 1165
1162PLOT <,6*S,U>,<5*S,-9*S,D>,<-10*S,,D>,<5*S,9*S,D>,<,-6*S,U>:
GOTO 1165
1163PLOT <4*S,4*S,U>,<-8*S,-8*S,D>,<,8*S,U>,<8*S,-8*S,D>,<-4*S,4
*S,U>:GOTO 1165
1164PLOT <,5*S,U>,<,-10*S,D>,<5*S,5*S,U>,<-10*S,,D>,<5*S,,U>
1165NEXT K
1170 GOSUB 1622
1180 P9=1:Y=Y1:X=X2:GOSUB 1200:Y=Y2:GOSUB 1200
1190 X=X1:GOSUB 1200:Y=Y1:GOSUB 1200
1195 GOSUB 1622:GOSUB 1270:RETURN
1200 X6=(X-X1)*X4-X5:Y6=(Y-Y1)*Y4-Y5
1210 X7=INT(ABS(X6)/999)+1:X8=INT(ABS(Y6)/999)+1
1220 IF X8<X7THEN 1230:X7=X8
1230 X6=INT(X6/X7):Y6=INT(Y6/X7)
1240 IF P9=1THEN 1250:PLOT X7<X6,Y6,U>:GOTO 1260
1250 PLOT X7<X6,Y6,D>
1260 X5=X5+X7*X6:Y5=Y5+X7*Y6:RETURN
1270 Y=Y1:P9=0:PLOT <1,,C>,<10,,S>:FOR J1=1TO 2
1280 FOR L=1 TO X2-X1
1290 L1=2*EXP(LOG(10)*(X1+L-1)):L2=L1
1300FOR J=1 TO 5
1320 X=LOG(L2)/LOG(10)
1340 GOSUB 1200
1350 PLOT <0,SGN(1.5-J1)*J*5,D>,<0,-SGN(1.5-J1)*J*5,D>
1360 L2=L2+L1:NEXT J:NEXT L:Y=Y2:NEXT J1
1380Y=Y1:SELECT PRINT 215
1390 FOR X=X1 TO X2:GOSUB 1200
1400 L1=EXP(LOG(10)*X)
1402 L2=INT(LOG(L1)/LOG(10))+1
1404 IF L2<1THEN 1410
1406 PLOT <-10*L2,-20,U>
1408 PRINT L1:PLOT <-10*2,20,U>:GOTO 1418
1410 PLOT <-10*(-1*L2+2),-20,U>
1412 CONVERT L1 TO STR(A$,1,-1*L2+3),(#.###########)
1414PRINT A$:A$=" "
1416 PLOT <-10*1,20,U>
1418 NEXT X
1430 L=(LOG(ABS(Y2-Y1)))/LOG(10)
1440 L1=EXP((L-INT(L))*LOG(10))
1450 IF L1>3THEN 1460:L1=0.25:GOTO 1470
1460 L1=1
1470 L1=L1*SGN(Y2-Y1)*10↑INT(L):L2=(INT(Y1/L1)+1)*L1
1480 Y=L2-L1:X=X1
1490 Y=Y+L1:IF SGN(Y2-Y1)*Y)=SGN(Y2-Y1)*Y2THEN 1510
1500 GOSUB 1200:PLOT <10,0,D>,<-10,0,D>:GOTO 1490
1510 Y=L2-L1:X=X2
1520 Y=Y+L1:IF SGN(Y2-Y1)*Y)=SGN(Y2-Y1)*Y2THEN 1540
1530 GOSUB 1200:PLOT <-10,0,D>,<10,0,D>:GOTO 1520
1540 Y=L2-L1:X=X1
1550 Y=Y+L1:IF SGN(Y2-Y1)*Y)=SGN(Y2-Y1)*Y2THEN 1585
1560 GOSUB 1200:PLOT <-10*(INT(LOG(Y)/LOG(10))+3),,U>
1570 PRINT Y:GOTO 1550
1585 Y=Y1:X=(X2+X1)/2:GOSUB 1200
1590 PLOT <-2*10,-40,U>:PRINT L$(L9)
1600 GOSUB 1622:SELECT PRINT 005:RETURN
1622PLOT <,,R>:X5,Y5=0:RETURN
1630 N0=0:REWIND :DATA SAVE OPEN "TOTAL":DATA SAVE N0
1640DATA SAVE END :DATA SAVE OPEN "ROT":REWIND :END
1700 PRINT "1.    CDNA-DNA BY HAP"
1710 PRINT "2.    CDNA-DNA BY S1"
```

```
1720 PRINT "3.    CDNA-RNA BY S1"
1730 INPUT "CHOOSE MODEL",W
1740 RETURN
1750 ON W GOTO 1755,1760,1765
1755 A(I0)=1/A(I0):RETURN
1760 A(I0)=(2↑(1/0.44)-1)/A(I0):RETURN
1765 A(I0)=LOG(2)/A(I0):RETURN
2000 P1=1000000:P2=-1:P3=1000:P4=-100
2020 FOR K=1 TO N
2030 IF P(K,1)>P1THEN 2040:P1=P(K,1)
2040 IF P(K,1)<P2THEN 2050:P2=P(K,1)
2050 IF P(K,2)>P3THEN 2060:P3=P(K,2)
2060 IF P(K,2)<P4THEN 2065:P4=P(K,2)
2065 NEXT K
2070 FOR I=1 TO M:I0=2*(I-1)+1
2080 A(I0)=EXP(LOG(P1)+(I-0.5)*((LOG(P2)-LOG(P1))/M))
2090 A1(I0)=1
2100 A(I0+1)=(P4-P3)/M:A1(I0+1)=1
2105 GOSUB 1750
2110 NEXT I
2120 A(2*M+1)=P3:A1(2*M+1)=1
2130 RETURN
```

Part II

```
5 DIM B(13,13),B1(13,13),G(13),W(13):L1=0.001:T=0:W0=1
10 MAT REDIM B(M0,M0),G(M0),W(M0),B1(M0,M0)
30 DIM F(7)
50 MAT W=ZER:MAT B=ZER:MAT G=ZER:MAT A=A0:S=0
60 FOR K=1TO N:KEYIN A$,555,555:X=P(K,1):Y=P(K,2):PRINT K,HEX(0C)

80 FOR L=1TO M+1:GOSUB 1000:Y=Y-V:F(L)=V:NEXT L:S=S+Y*Y
130 FOR J=1TO M0:IF D(J,2)>MTHEN 220:D0=(D(J,2)-1)*2+D(J,1)
160 ON D(J,1)GOTO 180,170
170 G(J)=F(D(J,2))/A(D0):GOTO 230
180 ON WGOTO 185,190,195
185 G(J)=A(D0+1)*X*(1-F(D(J,2))/A(D0+1))↑2:GOTO 230
190 G(J)=0.44*A(D0+1)*X*(1-F(D(J,2))/A(D0+1))↑(1.44/0.44):GOTO 23
0
195 G(J)=X*(A(D0+1)-F(D(J,2))):GOTO 230
220 G(J)=1:GOTO 230
230 W(J)=W(J)+Y*G(J)
240 FOR I=1TO J:B(I,J)=B(I,J)+G(I)*G(J):NEXT I:NEXT J:NEXT K
270 FOR J=1 TO M0:FOR I=1TO J:B(J,I)=B(I,J):NEXT I:NEXT J
285 PRINT SQR(S/N):MAT B1=B:GOTO 300
295 MAT B=B1
300 FOR J=1TO M0:B(J,J)=B(J,J)*(1+L1):NEXT J:MAT B=INV(B):MAT G=B
*W:MAT A=A0
320 J=1:FOR I=1TO M0
330 IF A1(J)=1THEN 350
340 J=J+1:GOTO 330
350 A(J)=A(J)+G(I):A2(J)=SQR(B(I,I))*SQR(S/(N-M0)):J=J+1:NEXT I
370 S1=0:FOR K=1TO N:X=P(K,1):Y=P(K,2)
380 FOR L=1TO M+1:GOSUB 1000:Y=Y-V:NEXT L:S1=S1+Y*Y:NEXT K
400 IF S>S1THEN 430:L1=L1*10:GOTO 295
430 MAT A0=A:L1=L1/10:T=T+1:PRINT "ITERATION",T
530 FOR L=1TO M:L0=(L-1)*2+1
531 GOSUB 1050:PRINTUSING 532,L$(L9)
532%          ###1/2 (STD.ERR)          % (STD.ERR)
533 PRINTUSING 534,L,R1,A2(L0)*R1/A(L0),A(L0+1),A2(L0+1)
534%##    ####.##### (#.##↑↑↑↑)    ###.### (#.##↑↑↑↑)
535 NEXT L
536 PRINTUSING 538,A(L0+2),A2(L0+2)
538%BACKGROUND=#####.##(##.###)
550 IF ABS((S-S1)/S)>0.000001THEN 50
555 S=S1
560 LOAD "PART 3"
1000 IF L<=MTHEN 1010:V=A(2*M+1):RETURN
1010 ON WGOTO 1020,1030,1040
1020 V=A(2*L)*(1-1/(1+P(K,1)*A(2*L-1))):RETURN
1030 V=A(2*L)*(1-1/((1+P(K,1)*A(2*L-1))↑0.44)):RETURN
1040 V=A(2*L)*(1-EXP(-P(K,1)*A(2*L-1))):RETURN
1050 ON WGOTO 1060,1070,1080
1060 R1=1/A(L0):RETURN
1070 R1=(2↑(1/0.44)-1)/A(L0):RETURN
1080 R1=LOG(2)/A(L0):RETURN
```

Part III

```
4 DIM N9$1
5SELECT PRINT 005
6 P9=0
10 INPUT "OUTPUT RESULTS ON TELETYPE",A$
20IF A$="N"THEN 540
30 SELECT PRINT 01D
40 PRINT :PRINT :PRINT B$:PRINT
45 PRINT "NO OF DATA POINTS=",N
50 PRINTUSING 60,T,W
60%RESULTS AFTER### ITERATIONS USING MODEL #
400 PRINT "ROOT MEAN SQUARE DIFF = ";
410 PRINT SQR(S1/N)
420 PRINT
425 PRINTUSING 430,L$(L9)
430 %           ###1/2 (STD.ERR)        % (STD.ERR)
530 FOR L=1TO M:L0=(L-1)*2+1:GOSUB 1100
532 PRINTUSING 534,L,R1,A2(L0)*R1/A(L0),A(L0+1),A2(L0+1)
534%## ######.##### (#.##↑↑↑↑)     ###.### (#.##↑↑↑↑)
535 NEXT L:PRINT
536 PRINTUSING 538,A(L0+2),A2(L0+2)
538%BACKGROUND=###.###% (##.###)
540 SELECT PRINT 005
560 INPUT "PLOT FITTED CURVE",A$
570 IF A$="N"THEN 610
575GOSUB 3100
580 P9=0:X=X1
585 R1=10↑X:Y=A(2*M+1)
587 FOR L=1 TO M:GOSUB 1000:Y=Y+F:NEXT L
590 GOSUB 3000:P9=1
595 X=X+(X2-X1)/100:IF X<=X2THEN 585
600 PLOT <,,U>,<,,R>:GOSUB 3100
610 INPUT "PLOT EACH COMPONENT",A$
620 IF A$<>"Y"THEN 710
650 Y=A(2*M+1)
660 FOR L=0 TO M
663 IF L<1THEN 670
664 IF L)1THEN 666
665 Y9=Y9+A(2*M+1):GOTO 670
666 Y9=Y9+A(2*(L-1))
670 P9=0:X=X1
675 R1=10↑X:IF L<1THEN 685
680 GOSUB 1000:Y=Y9+F
685 GOSUB 3000:P9=1
690 X=X+(X2-X1)/100:IF X<=X2THEN 675
700 NEXT L
710PLOT <,,U>,<,,R>:BACKSPACE 3F:LOAD
10000N WGOTO 1010,1020,1030
1010F=A(2*L)*(1-1/(1+R1*A(2*L-1))):RETURN
1020F=A(2*L)*(1-1/((1+R1*A(2*L-1))↑0.44)):RETURN
1030F=A(2*L)*(1-EXP(-R1*A(2*L-1))):RETURN
11000N WGOTO 1110,1120,1130
3060 X5=X5+X7*X6:Y5=Y5+X7*Y6:RETURN
3100PLOT <,,R>:X5,Y5=0:RETURN
3200 X7=INT(ABS(NO*26)/1000)+1
3210 PLOT X7<-NO*26/X7,0,U>,<,-50,U>
3220 RETURN
```

USE OF THE PROGRAM

This program operates in an interactive manner such that the user is prompted to enter the appropriate parameters. There are however a number of features which the user should be aware of before running the program. It is possible to run the program on previously entered data, referred to by the program as 'CURRENT DATA'. Alternatively the user can enter a new set of data each time the program is run. As indicated by the program, the data are entered as the co-ordinate pairs for each point ($C_0 t$ or $R_0 t$, % hybridised). Once data are entered, the user selects the number of components to be used in the fitting. The starting values for each component can be selected automatically by the program or can be entered by the user as $C_0 t_{1/2}$ and percentage. A background can also be used which can be held constant or allowed to vary. Once the program has starting values, the iterative procedure is used to improve them and should converge within $5-6$ iterations. If the result is unsatisfactory it may be necessary to re-run the program with a different number of components. This can be done without having to re-enter the data.

The plotter routines allow a visual presentation of the data points and the fitted curve. These routines may have to be adapted according to the type of computer used.

Array Storage

P(100,2)	Stores N data point co-ordinates
A(13) AO(13) A1(13) A2(13) D(13,2)	Stores parameter values for kinetic components during fitting procedure
B(13,13) B1(13,13) G(13) W(13)	Stores values for matrix inversion during each iteration

REFERENCES

1. Marquadt,D.W. (1963) *J. Soc. Ind. Appl. Math.*, **11**, 431.
2. Bevington,P.R. (1969) *Data Reduction and Error Analysis for the Physical Sciences*, McGraw-Hill Book Co., New York.
3. Kells,D.J.C. and Strauss,N.A. (1977) *Anal. Biochem.*, **80**, 344.
4. Pearson,W.R., Davidson,E.H. and Britten,R.J. (1977) *Nucleic Acids Res.*, **4**, 1727.

Suppliers of Specialist Items

The following list is not intended to be exhaustive but rather to give addresses of suppliers cited in this book. Many of the larger companies have subsidiaries in other countries whilst most of the smaller companies also market their products through agents. The name of a local supplier is most easily obtained by writing to the relevant address listed here, which is usually the head office.

Amersham International plc, White Lion Road, Amersham, Buckinghamshire, HP7 9LL, UK

Anglian Biotechnology Ltd., Unit 8, Hawkins Road, Colchester, Essex CO2 8JX, UK

BDH Chemicals Ltd., Broom Road, Poole, Dorset BH12 4NN, UK

Bethesda Research Laboratories Inc., P.O. Box 6009, Gaithersburg, MD 20877, USA

BioRad Laboratories Ltd., 2200 Wright Avenue, Richmond, CA 94804, USA
Also Caxton Way, Watford, Hertfordshire, UK

Boehringer Mannheim Biochemica, P.O. Box 31020, D-6800 Mannheim, FRG
Also Boehringer Mannheim House, Bell Lane, Lewes, East Sussex BN7 1LG, UK

Brand, Weirtheim, FRG

BRL; see Bethesda Research Laboratories

Digistrand Matra Optique SA, 37 Av. L. Brequet BP1, 78140 Velizy-Villacoublay, France

Dynatech Laboratories Ltd., Daux Road, Billingshurst, Sussex RH14 9SJ, UK

Eastman Kodak Co., DC Special Products Division, 2400 Mt. Read Blvd., Rochester NY 14650, USA

Edwards High Vacuum, Manor Royal, Crawley, West Sussex RH10 2ZA, UK

Falcon. Products available from Becton Dickinson U.K. Ltd., Between Towns Road, Cowley, Oxford   OX4 3LY, UK

Flow Laboratories Ltd., Dolphin House, Rockingham Road, Uxbridge, Middlesex UB8 2UE, UK

Flow Laboratories Inc., 7655 Old Springhouse Road, McLean, VA 22101, USA

Fullam, P.O. Box 444, Schenectady, NY 12301, USA

Gibco Ltd., P.O. Box 35, Trident House, Renfrew Road, Paisley PA3 4EF, UK

Gilson. Products available from Anachem Ltd., 15 Power Court, Luton, Bedfordshire LU1 3JJ, UK

Harelco, Gibbstown, NJ 08027, USA

Ilford, 14-22 Tottenham Street, London W1P OAH, UK

Kodak Ltd., Kodak House, Station Road, Hemel Hempstead, Hertfordshire HP1 1JU, UK

Life Sciences Inc., 2900 72nd Street North, St. Petersburg, FL 33710, USA
(products available from Anglian Biotechnology Ltd. and Northumbria
Biologicals Ltd. in the UK)

Matra S.A.; see Digistrand Matra Optique SA

Merck, D-6100 Darmstadt, FRG

Miles Laboratories, P.O. Box 37, Stoke Poges, Slough SL2 4LY, UK
Also P.O. Box 2000, Elkart, IN 46515, USA

Millipore, Ashby Road, Bedford, MA 01730, USA
Also 11-15 Peterborough Road, Harrow, Middlesex, UK

New England Biolabs Inc., 32 Tozer Road, Beverley, MA 01915, USA
Also products available from CP Laboratories Ltd, PO Box 22, Bishops
Stortford, Hertfordshire, UK

New England Nuclear, Postfach 401240, D-6072 Dreieich, FRG
Also 549 Albany Street, Boston, MA 02118, USA

NEN; see New England Nuclear

Northumbria Biologicals Ltd, S. Nelson Industrial Estate, Cramlington,
NE23 9HL, UK

PALL Corporation, 30 Sea Cliff Avenue, Glen Cove NY 11542, USA
Also PALL Process Filtration Ltd, Europa House, Havant Street, Portsmouth
PO1 3PD, UK

PCR Research Chemicals, Gainesville, Florida, USA

Pelanne Instruments, 77 bvd. Saint-Michel, F-75005 Paris, France

Pharmacia, S-75182 Uppsala, Sweden
Also Pharmacia House, Midsummer Boulevard, Milton Keynes, MK9 3HP, UK

Promega Biotech, 2800 S. Fish Hatchery Road, Madison, WI 53711, USA
Also products available from P and S Biochemicals, 38 Queensland Street,
Liverpool L7 3JG, UK

Razel Scientific Instruments Inc., 980 Hope Street, Stamford, CT 06907, USA

Sartorius Instruments Ltd., 18 Avenue Road, Belmont, Surrey SM2 6JD, UK

Sartorius GmbH, Postfach 19, D-3400, Göttingen, FRG

Schleicher and Schuell. Products available from Anderman & Co. Ltd., 145
London Road, Kingston-upon-Thames, Surrey KT2 6NH, UK

Sigma Chemical Co. Ltd., Fancy Road, Poole, Dorset BH17 7NH, UK
Also P.O. Box 14508, St. Louis, MO 63178, USA

Superior, Panel Marienfeld K.G., P.O. Box 1523, D-6990 Bad Mergentheim,
FRG

Waters Associates, Milford, MA 01757, USA

Whatman Ltd., Springfield Mill, Maidstone, Kent ME14 2LE, UK

Worthington Biochemical Corporation, Freehold, NJ 07728, USA

INDEX

Index

basic protocol, 133-135
hybridisation,
  kinetics, 84-85
  standard method, 133-135
  using high complexity probes, 136
  using low complexity probes, 136
  using cloned DNA, 135
  using genomic DNA, 135
SP6 polymerase,
  RNA probes, 40-42, 92, 185-187
Spreading of chromosomes, 188-190
Spreading of nucleic acids for electron
  microscopy,
    grids, 166
    protein monolayers, 169-172
    protocol, 170-172
    supporting films, 167-168
Stability of duplexes and hybrids,
  criterion, 9
  factors affecting,
    base composition, 9
    formamide, 80-81, 151, 164, 166
  supporting films, 167-168
    fragment length, 9
    ionic strength, 9, 80-81
    sequence mismatch, 8-9
  in electron microscopy, 176-177
  of mismatched sequences, 81
  of perfectly matched sequences, 80-81
  temperature, 166
  T_m as measure of, 4, 80
Staining,
  nucleic acids for electron microscopy,
    172
  tissue sections, 200-202
Standards,
  autoradiography, 100
  densitometry, 100
  electron microscopy, 173
  kinetic, 10, 59-62, 71
Strand separation, 147, 153-154
  see also Denaturation
Stringency,
  choice of hybridisation conditions, 93-94,
    105-106
  determination of, 119
  discrimination between related sequences,
    81-82
  factors involved, 78-79, 119
  in in situ hybridisation, 192
  in screening libraries, 119
  see also Criterion
Supporting films for electron microscopy,
  167

Temperature of hybridisation,
  effect on,
    hybrid stability, 166
    rate, 65-66, 78
  for in situ hybridisation, 192
  for primer extension, 154-156
  for screening libraries, 119
  lowering by formamide, 106
  optimum, 6, 105-106
Terminal deoxynucleotidyl transferase,
  for probe end-labelling, 36-37
Time of hybridisation,
  optimum, 82-86, 106-107
Tissue,
  acetylation, 192, 196
  DNA isolation from, 21-22
  denaturation of DNA in, 192
  embedding, 194-195
  fixation, 188-189, 194-196
  freeze substitution, 194-195
  preparation for in situ hybridisation,
    examples, 194-196
    for chromosome spreads, 188-190
    general considerations, 187-188,
      193-194
  protease treatment, 196
  sectioning, 194-196
  staining, 200-202
T_m,
  definition, 4
  factors affecting, 8-9, 64-65, 78, 80-81
  formamide, effect on, 64-65, 78, 80-81
  hybrid stability, 4, 80-81
  in electron microscopy, 163-165
  measurement of, 107-108
  of mismatched sequences, 81
  of perfectly matched sequences, 80-81
Transcripts, RNA,
  heterogeneity, 156-158
  mapping,
    nuclease S1, 143-152
    primer extension, 152-159
  sequencing of primer extension product,
    153, 156-158
  splice junctions, location of, 143-152

Uranyl acetate staining in electron
  microscopy, 172

Vectors, choice for cloning, 22-24
Viscosity,
  effect on hybridisation rate, 64, 80

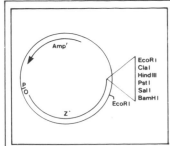

DNA cloning
(Volumes I and II)
a practical approach

Edited by D M Glover,
Imperial College of Science and
Technology, London

Published
in the
Practical
Approach
series

A STEP-BY-STEP GUIDE TO PROVEN NEW TECHNIQUES

Breakthroughs in the manipulation of DNA have already revolutionised biology; they are set to do the same for drug and food production. *DNA cloning* contains the background and detailed protocols for molecular biologists to perform these experiments with success. It supersedes previous manuals in describing recent developments with widespread applications that use *E coli* as the host organism.

Up-to-the-minute contributions cover the use of bacteriophage λ insertion vectors for cDNA cloning and the use of phage λ replacement vector systems to select recombinants for DNA cloning.

Two chapters evaluate *E coli* transformation and methods for *in vitro* mutagenesis of DNA cloning in other organisms including yeast, plant cells and Gram-negative and Gram-positive bacteria. Finally, the last three chapters of Volume II offer three different approaches to the introduction of cloned genes into animal cells.

Contents
Volume I

The use of phage lambda replacement vectors in the construction of representative genomic DNA libraries *K Kaiser and N E Murray* ● Constructing and screening cDNA libraries in λ gt10 and λ gt11 *T V Huynh, R A Young and R W Davis* ● An alternative procedure for synthesising double-stranded cDNA for cloning in page and plasmid vectors *C Watson and J F Jackson* ● Immunological detection of chimeric β-galactosidases expressed by plasmid vectors *M Koenen, H W Gresser and B Muller-Hill* ● The pEMBL family of single-stranded vectors *L Dente, M Sollazzo, C Baldari, G Cesareni and R Cortese* ● Techniques for transformation of *E coli* *D Hanahan* ● The use of genetic markers for the selection and the allelic exchange of *in vitro* induced mutations that do not have a phenotype in *E coli G Cesareni, C Traboni, G Ciliberto, L Dente and R Cortese* ● The oligonucleotide-directed construction of mutations in recombinant filamentous phage *H-J Fritz* ● Broad host range cloning vectors for Gram-negative bacteria *F C H Franklin* ● Index

Volume II

Bacillus cloning methods *K G Hardy* ● Gene cloning in *Streptomyces I S Hunter* ● Cloning in yeast *R Rothstein* ● Genetic engineering of plants *C P Lichtenstein and J Draper* ● P element-mediated germ line transformation of *Drosophila R Karess* ● High-efficiency gene transfer into mammalian cells *C Gorman* ● The construction and characterisation of vaccinia virus recombinants expressing foreign genes *M Mackett, G L Smith and B Moss* ● Bovine papillomavirus DNA: an eukaryotic cloning vector *M S Campo* ● Index

Volume I: *June 1985; 204pp;*
0 947946 18 7 (softbound)
Volume II: *June 1985; 260pp;*
0 947946 19 5 (softbound)
Volumes I and II; *0 947946 20 9*

For details of price
and ordering consult
our current catalogue
or contact:

IRL Press Ltd,
Box 1, Eynsham,
Oxford OX8 1JJ, UK

IRL Press Inc,
PO Box Q,
McLean VA 22101,
USA

◇ **IRL PRESS**
Oxford · Washington DC